Manual of Nerve Conduction Studies

Manual of Nerve Conduction Studies

Ralph M. Buschbacher, M.D.

Demos

Demos Medical Publishing, Inc., 386 Park Avenue South, New York, New York 10016

© 2000 by Demos Medical Publishing, Inc. All rights reserved. This book is protected by copyright. No part of it may be reproduced, stored in a retrieval system, or transmitted in any form or by any means, electronic, mechanical, photocopying, recording, or otherwise, without the prior written permission of the publisher.

Library of Congress Cataloging-in-Publication Data

Available from the publisher upon request

Printed in Canada

Dedicated to the Memory of Rick DeCarlo, M.D.

Rick DeCarlo was in residency with me. We spent our first rotation working side by side, and so came to think of ourselves as partners. Rick told me that he had had multiple myeloma in medical school. He was thought to be cured, but he wasn't. He died within a year after we finished residency. I am never sad when I think of Rick. I just think of the good times we had together and smile. So I dedicate this book to my good friend Rick DeCarlo.

To
Lois
Michael
Peter
John
Walter

Contents

Preface   xi

Introduction   xiii

1. Upper Extremity/Cervical Plexus/Brachial Plexus Motor Studies   1

   Axillary motor nerve to the deltoid   2

   Long thoracic motor nerve to the serratus anterior   6

   Median Nerve
      Median motor nerve to the abductor pollicis brevis   10
      Median motor nerve to the flexor carpi radialis, including H-reflex   18
      Median motor nerve (anterior interosseous branch) to the flexor pollicis longus   24
      Median motor nerve (anterior interosseous branch) to the pronator quadratus   28
      Median motor nerve to the 1st lumbrical   32
      Median motor nerve to the 2nd lumbrical   36

   Musculocutaneous motor nerve to the biceps brachii   40

   Phrenic motor nerve to the diaphragm   44

   Radial Nerve
      Radial (posterior interosseus) motor nerve to the extensor carpi ulnaris and brachioradialis   50
      Radial motor nerve to the extensor digitorum communis   54
      Radial motor nerve to the extensor indicis proprius   58

   Suprascapular motor nerve to the supraspinatus and infraspinatus   64

Thoracodorsal motor nerve to the latissimus dorsi	70
Ulnar Nerve	
Ulnar motor nerve to the abductor digiti minimi	74
Ulnar motor nerve to the palmar interosseous	82
Ulnar motor nerve to the 1st dorsal interosseous	86

2. Upper Extremity Sensory and Mixed Studies — 91

Lateral antebrachial cutaneous sensory nerve study	92
Medial antebrachial cutaneous sensory nerve study	98
Median Nerve	
Median sensory nerve to the 2nd and 3rd digits	104
Median palmar cutaneous sensory nerve study	112
Posterior antebrachial cutaneous sensory nerve study	116
Radial sensory nerve study to the base of the thumb	120
Ulnar Nerve	
Ulnar dorsal cutaneous sensory nerve study	124
Ulnar sensory nerve study to the 5th digit	130
Comparative Studies	
Median and radial sensory nerves to 1st digit	138
Median and ulnar mixed nerve studies	142
Median and ulnar sensory studies to the 4th digit	148

3. Lower Extremity Motor Nerves — 153

Femoral motor nerve to the quadriceps	154
Peroneal Nerve	
Peroneal motor nerve to the extensor digitorum brevis	158
Peroneal motor nerve to the peroneus brevis	164
Peroneal motor nerve to the peroneus longus	168
Peroneal motor nerve to the tibialis anterior	172
Sciatic nerve	176
Tibial Nerve	
Tibial motor nerve (inferior calcaneal branch) to the abductor digiti minimi	180

Tibial motor nerve (medial plantar branch) to the
abductor hallucis 184
Tibial motor nerve (lateral plantar branch) to the
flexor digiti minimi brevis 190
H-reflex to the calf 194

4. Lower Extremity Sensory and Mixed Studies 199

Lateral femoral cutaneous sensory study 200

Medial calcaneal sensory study 206

Medial femoral cutaneous sensory study 210

Peroneal Nerve
   Deep peroneal sensory study 214
   Superficial peroneal sensory study (medial and
      intermediate dorsal cutaneous branches) 218

Posterior femoral cutaneous sensory study 224

Saphenous Nerve
   Saphenous sensory study (distal technique) 228
   Saphenous sensory study (proximal technique) 232

Sural Nerve
   Sural lateral dorsal cutaneous branch sensory study 236
   Sural sensory study 240

Tibial nerve compound action potentials (medial and
lateral plantar nerves) 244

5. Cranial Nerves 249

Blink reflex 250
Cranial nerve VII 256
Cranial nerve XI 260

6. Root Stimulation 263

Cervical nerve root stimulation 264
Lumbar root stimulation 268

7. **Other Studies of Interest** 273

   Appendix 279

   Index 283

Preface

It is often said that each EMG laboratory should develop its own set of normal values. Yet in reality such a proposition is unreasonable for most facilities, and most electromyographers utilize, at least to some extent, numbers generated by others—hence the need for a reference book such as this. One frustration for me has always been knowing which numbers to rely on, especially given the rather wide range of protocols, techniques, filter setting, measurements, and so forth quoted in the literature. My goal—first for myself and later for this book—was to collect a set of nerve conduction tests that I could trust.

Many of the commonly used nerve conduction study techniques and reference values rely on data generated years ago, which is not necessarily a bad thing except that in the pioneering days of electromyography there were a lot of unknowns—things such as the effect of temperature, age, and height on normal values. The early studies often included small numbers of volunteers without controlling for these factors. And often these classic studies have continued to be used as the "gold standards."

I have redone some of the common tests, using modern protocols and a large and varied subject group. The goal was to derive a set of normal reference values, not just for the summated population, but for each particular patient being tested, depending on his or her demographic characteristics. When the literature has offered well-documented techniques with good protocols, I have also included them in this manual. Other tests obviously need to be updated as well, and more work needs to be done on improving our "normals." In toto, however, this manual should provide a fairly comprehensive, up-to-date set of reference values for clinical use by the electromyographer. I hope that the reader will benefit from my efforts and will truly use this handbook at the EMG bedside.

I thank TECA Corporation and XLTEK Ltd. for their help in supporting the research to derive the data in some of the cited studies.

Introduction

To best appreciate and optimally use this handbook, the reader should understand its intended uses and limitations. The book is intended to be a bedside reference in which to quickly look up how to perform a technique as a refresher or to look up reference values that have not been committed to memory. It is *not* intended to be a comprehensive text or even a teaching manual. It is presupposed that the reader has a good working knowledge of how to perform nerve conduction studies. If you are an inexperienced practitioner, this is not the book for you—yet. You might consider first reading one of the many available excellent textbooks.

Some of the more commonly used techniques included in this manual were studied by me as part of a large investigation on nerve conduction studies. The primary goal of the project was to develop a more accurate database for the general population. The subjects were recruited to represent a large age range, with both genders and both blacks and whites included. Further demographic and physical characteristics were recorded so that a set of normal reference values could be generated, not just for the population at large, but for specific subsets of age, sex, height, and so forth. This database is presented so that the reader can look up the normal range for, say, a 52-year-old obese female or a 21-year-old thin male. The only factors that seemed to affect the normal ranges were in some cases sex, body mass index (BMI)—kg/m^2, and height. Where applicable, the normal ranges are divided into subsets based on these characteristics. For sex and height it obviously is easy to determine into which category a given patient fits. For BMI, the Appendix provides a quick reference for looking up the applicable value. Alternatively, the practitioner may perform his own calculation to derive the BMI of a given patient.

For studies that were not done by me, I have tried to include techniques that are well accepted and reliable. In some cases, in which there was more than one technique that could be done, I have included descriptions from more than one author. I have also included other references and alternate techniques under the heading "Additional Readings/Alternate Techniques."

Another goal for my study and for this book was to better codify the acceptable differences between nerves of the same limb or the opposite limb. This information is included under the "Acceptable Differences" section of the applicable tests.

The studies that I performed were done under "real-life" situations. Thus, bar electrodes were used for many sensory tests, despite the fact that optimal amplitude may be obtained with a slightly greater electrode separation. This was done so that the practitioner can compare his results with those presented in this book, without deviating from standard practice. In a similar vein, ground electrodes were placed where most people put them—on the back of the hand. Temperatures were adjusted by warming with a hair dryer if necessary; correction factors were not used. Skin was specially prepared or abraded only if necessary (extremely rarely). Needle recording was avoided in order to be able to use amplitude data, but when it is necessary, the appropriate references are cited. Similarly, needle stimulation was avoided. Unless otherwise stated, all studies in this book use both surface stimulation and recording. For some techniques needle stimulation may be needed in some patients, especially in the obese or in people whose skin is especially thick.

Several authors have noted that normative data for nerve conduction studies do not necessarily follow a normal (Gaussian) distribution. The data often need to be converted by a logarithmic, square root, or other transformation before creating a mean ± 2 S.D. range of normal; this was done as needed for the studies that I performed. Using percentiles of normal is also an acceptable substitute, so for the nerve studies that I performed, I include normal ranges defined as both mean ± 2 S.D. and 97th (3rd) percentile of observed values. For the studies of other authors, I include the results and normal ranges that were derived by them.

CHAPTER 1

Upper Extremity/
Cervical Plexus/
Brachial Plexus
Motor Studies

AXILLARY MOTOR NERVE TO THE DELTOID

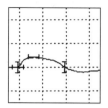

Typical waveform appearance

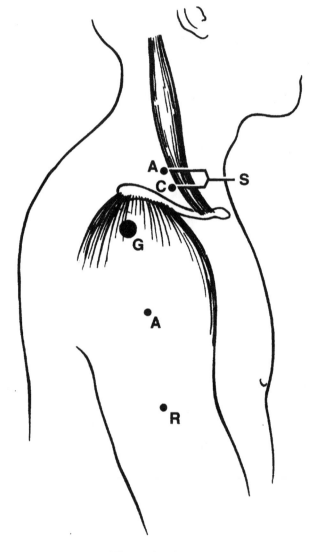

Electrode Placement

Active electrode (A): Placement is over the most prominent portion of the middle deltoid.

Reference electrode (R): Placement is over the junction of the deltoid muscle and its tendon of insertion.

Ground electrode (G): Placement is over the posterolateral shoulder.

Stimulation point (S): Erb's point—the cathode (C) is placed slightly above the upper margin of the clavicle lateral to the clavicular head of the sternocleidomastoid muscle. The anode (A) is superomedial.

Machine settings: Standard motor settings are used.

Nerve fibers tested: C5 and C6 nerve roots, through the upper trunk, posterior division, and posterior cord of the brachial plexus.

Normal values (1) (62 subjects) (room temperature 75–84 degrees Fahrenheit):

Onset latency (msec)

Mean	S.D.	Normal Range	Observed Range
3.9	0.5	2.8–5.0	2.9–5.2

Helpful Hints

- The active electrode is placed over the region of greatest muscle mass, localized upon abduction of the shoulder.
- The distance between the stimulation point and the active electrode ranges from approximately 14.8 to 21.0 cm, measured with obstetric calipers with the arm by the side.

4 Upper Extremity/Cervical Plexus/Brachial Plexus Motor Studies

Notes

REFERENCE

1. Kraft GH: Axillary, musculocutaneous and suprascapular nerve latency studies. *Arch Phys Med Rehabil* 1972; 53:383–387.

ADDITIONAL READING/ALTERNATE TECHNIQUE

1. Gassel MM: A test of nerve conduction to muscles of the shoulder girdle as an aid in the diagnosis of proximal neurogenic and muscular disease. *J Neurol Neurosurg Psychiatry* 1964; 27:200–205.

LONG THORACIC MOTOR NERVE TO THE SERRATUS ANTERIOR

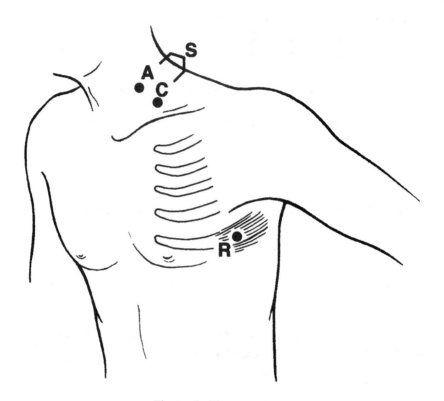

Electrode Placement

Recording electrodes: A concentric needle electrode (R) is placed at the digitation of the serratus anterior along the midaxillary line over the 5th rib. (1) Alternately, a monopolar needle electrode can be placed in this same site, with the reference electrode 2 cm caudal and the ground electrode at the anterior axillary line over the 12th rib level (2).

Stimulation point (S): Erb's point—the cathode (C) is placed slightly above the upper margin of the clavicle lateral to the clavicular head of the sternocleidomastoid muscle. The anode (A) is superomedial.

Machine settings: Standard motor settings are used.

Nerve fibers tested: Anterior primary branches of the C5, C6, and C7 nerve roots.

Normal values:

Onset latency (1) (msec—44 subjects, concentric needle recording)

Age (years)	Mean	S.D.	Mean + 2 S.D.
20–35	3.6	0.3	4.2
36–50	3.8	0.4	4.4
51–65	4.0	0.4	4.8

Onset latency (2) (msec—25 subjects, monopolar needle recording, room temperature 21–23 degrees Celsius)

Mean	S.D.	Mean + 2 S.D.
3.9	0.6	5.1

Helpful Hints

- In the study by Alfonzi and colleagues (1) the mean distance between stimulating and recording electrodes was 23.6 ± 1 cm, measured with obstetric calipers (range 22–25 cm). The latency was found to correlate with distance, with approximately a 0.2

msec increase in latency for each 1 cm increase in distance. In Kaplan's study (2) the distances ranged from 17 to 23 cm.

- Surface recording techniques have also been described (1,3,4). Ma and Liveson (3) studied 15 subjects and placed the active electrode over the midaxillary line of the 5th or 6th ribs with the reference electrode at the anterior axillary line of the same rib and reported a latency of 3.0 ± 0.2 msec. Alfonzi and coworkers (1) utilized surface recording with the active electrode at the digitation of the serratus anterior along the midaxillary line of the 5th rib with the reference electrode 3 cm in front of the active electrode. They reported latencies of 3.2 ± 0.3 msec, 3.3 ± 0.3 msec, and 3.3 ± 0.3 msec, and amplitudes of 4.3 ± 3.0 mV, 3.8 ± 2.4 mV, and 2.7 ± 1.2 mV for the age groups of 20–35, 36–50, and 51–65 years. They concluded that these recordings may be contaminated by volume conduction from other muscles and recommended using a needle recording technique. Cherrington (4) also studied this nerve using surface recording in 20 normal subjects. Stimulation was applied at Erb's point and recording was just lateral to the nipple. Normal latencies ranged from 2.6 to 4.0 msec over a distance of 18.0–22.0 cm.

- Proper needle placement can be confirmed with active protraction.

- If the recording electrode is placed too far posteriorly, it may result in erroneous recording from the latissimus dorsi.

Notes

REFERENCES

1. Alfonsi E, Moglia A, Sandrini G, et al: Electrophysiological study of long thoracic nerve conduction in normal subjects. *Electromyogr clin Neurophysiol* 1986; 26:63–67.
2. Kaplan PE: Electrodiagnostic confirmation of long thoracic palsy. *J Neurol Neurosurg Psychiatry* 1980; 43:50–52.
3. Ma DM, Liveson JA: *Nerve conduction handbook.* Philadelphia: FA Davis, 1983.
4. Cherrington M: Long thoracic nerve: conduction studies. *Diseases of the Nervous System* 1972; 33:49–51.

ADDITIONAL READING/ALTERNATE TECHNIQUE

1. LoMonaco M, DiPasqua PG, Tonali P: Conduction studies along the accessory, long thoracic, dorsal scapular, and thoracodorsal nerves. *Acta Neurol Scand* 1983; 68:171–176.

MEDIAN NERVE

MEDIAN MOTOR NERVE TO THE ABDUCTOR POLLICIS BREVIS

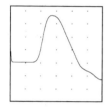

Typical waveform appearance

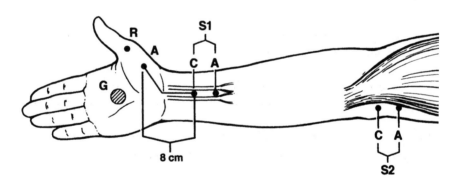

Electrode Placement

Active electrode (A): Placement is halfway between the midpoint of the distal wrist crease and the first metacarpophalangeal joint.

Reference electrode (R): Placement is slightly distal to the first metacarpophalangeal joint.

Ground electrode (G): Placement is on the dorsum of the hand. If stimulus artifact interferes with the recording, the ground may be placed near the active electrode, between this electrode and the cathode.

Stimulation point 1 (S1): The cathode (C) is placed 8 cm proximal to the active electrode, in a line measured first to the midpoint of the distal wrist crease and then to a point slightly ulnar to the tendon of the flexor carpi radialis. The anode (A) is proximal.

Stimulation point 2 (S2): The cathode (C) is placed slightly medial to the brachial artery pulse in the antecubital region. The anode (A) is proximal.

F-wave stimulation: The cathode (C) is positioned as for stimulation point 1, but with the anode distal.

Machine settings: Sensitivity—5 mV/division, Low frequency filter—2–3 Hz, High frequency filter—10 kHz, Sweep speed—2 msec/division

Nerve fibers tested: C8 and T1 nerve roots, through the lower trunk, anterior division, and medial cord of the brachial plexus.

Upper Extremity/Cervical Plexus/Brachial Plexus Motor Studies

Normal values (1) (249 subjects) (skin temperature over the dorsum of the hand greater than or equal to 32 degrees Celsius):

Onset latency (msec)

Males

Age Range	Mean	S.D.	Mean + 2 S.D.	97th Percentile
19–49	3.8	0.4	4.6	4.6
50–79	4.0	0.4	4.8	4.7

Females

Age Range	Mean	S.D.	Mean + 2 S.D.	97th Percentile
19–49	3.5	0.4	4.3	4.4
50–79	3.8	0.4	4.6	4.4
All subjects	3.7	0.5	4.7	4.5

Amplitude (mV)

Age Range	Mean	S.D.	Mean − 2 S.D.	3rd Percentile
19–39	11.9	3.6	4.7	5.9
40–59	9.8	2.8	4.2	4.2
60–79	7.0	2.6	1.8	3.8
All subjects	10.2	3.6	3.0	4.1

Area of negative phase (µVsec)

Age Range	Mean	S.D.	Mean − 2 S.D.	3rd Percentile
19–49	37.4	12.9	11.6	14.6
50–59	30.9	8.6	13.7	15.3
60–79	23.7	9.3	5.1	11.9
All subjects	33.7	12.8	8.1	12.4

Duration of negative phase (msec)

Age range	Mean	S.D.	Mean + 2 S.D.	97th Percentile
19–79	5.9	0.9	7.7	8.0

Nerve conduction velocity (m/sec)

Males

Age range	Mean	S.D.	Mean − 2 S.D.	3rd Percentile
19–39	58	4	50	49
40–79	55	5	45	47

Females

Age range	Mean	S.D.	Mean − 2 S.D.	3rd Percentile
19–39	60	3	54	53
40–79	57	5	47	51
All subjects	57	5	47	49

F-wave latencies (msec) (2) (195 subjects)—shortest of 10 stimuli

Age 19–49

Height in cm (in.)	Mean	S.D.	Mean + 2 S.D.	97th Percentile
< 160 (5′3″)	23.7	1.0	25.7	
160–169 (5′3″–5′6″)	25.3	1.6	28.5	
170–179 (5′7″–5′10″)	27.3	1.8	30.9	
≥ 180 (5′11″)	28.9	2.3	33.5	

Age 50–79

Height in cm (in.)	Mean	S.D.	Mean + 2 S.D.	97th Percentile
< 160 (5′3″)	25.2	1.7	28.6	
160–169 (5′3″–5′6″)	27.5	1.4	30.3	
170–179 (5′7″–5′10″)	28.7	1.4	31.5	
≥ 180 (5′11″)	30.4	1.9	34.2	
All subjects	26.8	2.4	31.6	31.4

Acceptable Differences

The upper limit of normal increase in latency from one side to the other is 0.7 msec.

The upper limit of normal decrease in amplitude from one side to the other is 54%.

The upper limit of normal decrease in nerve conduction velocity from one side to the other is 9 m/sec.

The upper limit of normal decrease in amplitude from wrist to elbow stimulation is 24%.

The upper limit of normal side to side difference in the shortest F-wave latency is 2.2 msec.

The upper limit of normal increase between median and ulnar latency in the same hand (ulnar recording from the abductor digiti minimi) has been reported to be 1.0 msec, although for this determination anatomic landmarks and not measurements were used to localize the stimulation sites (3).

Helpful Hints

- Care should be taken to not concomitantly stimulate the ulnar nerve. The direction of thumb twitch will help in making sure that only the median nerve is stimulated. The waveforms, especially the deflections from baseline, should be similar on proximal and distal stimulation.

- Stimulation can also be performed at the palm. If the amplitude with palm stimulation is significantly greater than with wrist stimulation, this can be a sign of neurapraxia at the wrist. Pease and coworkers (4) showed that the increase in amplitude with wrist stimulation is significantly larger in persons with carpal tunnel syndrome than in normal controls. Proximal to distal amplitude ratios of 0.5–0.8 have been recommended as the limits of normality (5,6). The 0.5 value is more conservative and seems reasonable for clinical use.

    Palmar stimulation may cause a direct excitation of the thenar muscle mass or of the deep branch of the ulnar nerve. It is helpful to move the cathode slightly distally on the palm and stimulate the patient a few times while repositioning the cathode gradually more proximal to optimize the resultant waveform recording. Stimulus artifact can be a problem and may be minimized by rotating the anode about the cathode and stimulating at various locations. Because the skin of the palm is thick, a longer pulse duration may be needed. Needle stimulation may be necessary in some cases.

- More proximal stimulation can also be performed at the axilla and at Erb's point in the supraclavicular fossa. This can allow determination of waveform changes across more proximal segments of the nerve and calculation of more proximal nerve conduction velocity. When calculating the conduction velocity of the Erb's point-to-axilla segment, obstetric calipers are used to measure the distance.

- Anomalous innervation due to a Martin-Gruber (median to ulnar) anastomosis in the forearm is common, although it is much less commonly clinically significant during electrodiagnostic studies. When present in a patient with carpal tunnel syndrome, it may cause confusion. For instance, a complete block of the median nerve to wrist stimulation may seem to be reversed on elbow stimulation. Martin-Gruber anastomosis should be suspected if the median motor amplitude is larger on elbow stimulation than on wrist stimulation, and in persons with median nerve slowing across the wrist who have a higher than normal conduction velocity across the forearm. It should also be suspected if proximal (but not distal) median nerve stimulation results in an initially positive deflection.

Martin-Gruber anastomosis can usually be confirmed by repositioning the active electrode to the first dorsal interosseous muscle. Stimulation of the median nerve at the elbow, but not the wrist, results in a negative deflection. Stimulation at the elbow should also result in a significantly larger amplitude response than with wrist stimulation (7). An accurate forearm conduction velocity cannot be calculated in the person with carpal tunnel syndrome and a Martin-Gruber anastomosis.

Notes

REFERENCES

1. Buschbacher RM: Median nerve motor conduction to the abductor pollicis brevis. *Am J Phys Med Rehabil* 1999; 78:S1–S8.
2. Buschbacher RM: Median nerve F-waves. *Am J Phys Med Rehabil* 1999; 78:S32–S37.
3. Felsenthal G: Median and ulnar distal motor and sensory latencies in the same normal subject. *Arch Phys Med Rehabil* 1977; 58:297–302.
4. Pease WS, Cunningham ML, Walsh WE, Johnson EW: Determining neurapraxia in carpal tunnel syndrome. *Am J Phys Med Rehabil* 1988; 67:117–119.
5. Ross MA, Kimura J: AAEM case report #2: the carpal tunnel syndrome. *Muscle Nerve* 1995; 18:567–573.
6. Fitz WR, Mysiw J, Johnson EW: First lumbrical latency and amplitude: control values and findings in carpal tunnel syndrome. *Am J Phys Med Rehabil* 1990; 69:198–201.
7. Sun SF, Streib EW: Martin-Gruber anastomosis: electromyographic studies. *Electromyogr clin Neurophysiol* 1983; 23:271–285.

ADDITIONAL READINGS/ALTERNATE TECHNIQUES

1. Melvin JL, Harris DH, Johnson EW: Sensory and motor conduction velocities in the ulnar and median nerves. *Arch Phys Med Rehabil* 1966; 47:511–519.
2. Melvin JL, Schuchmann JA, Lanese RR: Diagnostic specificity of motor and sensory nerve conduction variables in the carpal tunnel syndrome. *Arch Phys Med Rehabil* 1973; 54:69–74.
3. Perez MC, Sosa A, Acevedo CEL: Nerve conduction velocities: normal values for median and ulnar nerves. *Bol Asoc Med P Rico* 1986; 78:191–196.
4. Falco FJE, Hennessey WJ, Braddom RL, Goldberg G: Standardized nerve conduction studies in the upper limb of the healthy elderly. *Am J Phys Med Rehabil* 1992; 71:263–271.
5. Hennessey WJ, Falco FJE, Braddom RL: Median and ulnar nerve conduction studies: normative data for young adults. *Arch Phys Med Rehabil* 1994; 75:259–264.

MEDIAN MOTOR NERVE TO THE FLEXOR CARPI RADIALIS, INCLUDING H-REFLEX (1)

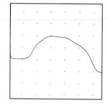
Typical waveform appearance

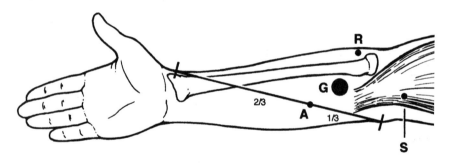
Electrode Placement

Active electrode (A): Placement is over the belly of the flexor carpi radialis, usually one third of the distance from the medial epicondyle to the radial styloid.

Reference electrode (R): Placement is over the brachioradialis.

Ground electrode (G): Placement is between the stimulating and recording electrodes.

Stimulation point (S): The median nerve is stimulated at the elbow with a 0.5–1.0 msec rectangular pulse with a frequency not more than 0.5 Hz. The cathode is proximal, and the anode is distal.

Machine settings: Standard motor settings are used, with a sweep speed of 5 msec/division and a sensitivity of 500 µV/division (1,2).

Nerve fibers tested: C6, C7, and C8 nerve roots, through the upper, middle, and lower trunks, anterior divisions, and the medial and lateral cords of the brachial plexus.

Normal values (1) (39 subjects) (room temperature of 70 degrees Fahrenheit):

M-wave (measured from same recording as H-reflex)

Onset latency (msec)

Mean	S.D.	Mean + 2 S.D.
3.0	0.5	4.0

Amplitude (mV)

Mean	S.D.	Mean − 2 S.D.
7.6	2.5	2.6

H-reflex

Onset latency (msec)

Mean	S.D.	Mean + 2 S.D.
15.9	1.5	18.9

Amplitude (mV, baseline to highest negative peak)

Mean	S.D.	Mean − 2 S.D.
1.6	0.4	0.8

Acceptable Difference

The upper limit of normal side to side difference in H-reflex latency is 1.0 msec (mean + 2 S.D.).

Helpful Hints

- The paper's author states that to be accepted as an H-reflex, the response must be obtained either without an M-response or with only a small M-response preceding it; its latency must be shortened when the nerve is stimulated proximally, and its amplitude must be decreased with increasing stimulation frequency. (In the opinion of this book's author, it may be difficult to obtain an H-reflex amplitude greater than the M-wave amplitude.)
- When using these criteria, 90% of normal subjects were found to have an elicitable H-reflex. The author reports that in none of his subjects was the H-reflex absent on only one side, but this must be interpreted with caution because of the small sample size.
- An alternate description places the reference electrode over the distal tendinous area of the forearm. The muscle can be palpated just medial to the pronator teres and can be palpated with resisted wrist flexion (2).
- In the paper describing the above-referenced values, the cathode was placed proximally for determining both the H-reflex and the M-wave latencies. M-wave recordings usually are made with the anode proximal.

- Kraft and Johnson (2) report that H-reflex latency is 17 ± 1.7 msec with highly variable amplitude and an upper limit of normal side to side difference of 0.85 msec.
- Facilitation may be necessary to obtain an H-reflex response. The elbow should be slightly flexed.
- With supramaximal stimulation, an F-wave response may be mistaken for an H-reflex.

Notes

REFERENCES

1. Jabre JF: Surface recording of the H-reflex of the flexor carpi radialis. *Muscle Nerve* 1981; 4:435–438.
2. Kraft GH, Johnson EW: Proximal motor nerve conduction and late responses: an AAEM workshop. American Association of Electrodiagnostic Medicine, Rochester, Minnesota, 1986.

ADDITIONAL READINGS/ALTERNATE TECHNIQUES

1. Deschuytere J, Rosselle N, De Keyser C: Monosynaptic reflexes in the superficial forearm flexors in man and their clinical significance. *J Neurol Neurosurg Psychiatry* 1976; 39:555–565.
2. Garcia HA, Fisher MA, Gilai A: H reflex analysis of segmental reflex excitability in flexor and extensor muscles. *Neurology* 1979; 29:984–991.
3. Ongerboer De Visser BW, Schimsheimer RJ, Hart AAM: The H-reflex of the flexor carpi radialis muscle: a study in controls and radiation-induced brachial plexus lesions. *J Neurol Neurosurg Psychiatry* 1984; 47:1098–1101.
4. Schimsheimer RJ, Ongerboer De Visser BW, Kemp B: The flexor carpi radialis H-reflex in lesions of the sixth and seventh cervical nerve roots. *J Neurol Neurosurg Psychiatry* 1985; 48:445–449.

5. Schimsheimer RJ, Ongerboer De Visser BW, Kemp B, Bour LJ: The flexor carpi radialis H-reflex in polyneuropathy: relations to conduction velocities of the median nerve and the soleus H-reflex latency. *J Neurol Neurosurg Psychiatry* 1987; 50:447–452.
6. Miller TA, Newall AR, Jackson DA: H-reflexes in the upper extremity and the effects of voluntary contraction. *Electromyogr clin Neurophysiol* 1995; 35:121–128.

MEDIAN MOTOR NERVE (ANTERIOR INTEROSSEOUS BRANCH) TO THE FLEXOR POLLICIS LONGUS

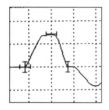

Typical waveform appearance

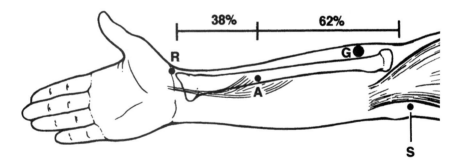

Electrode Placement

Active electrode (A): Placement is on the lateral forearm, 38% of the distance from the distal volar crease of the wrist to the antecubital crease of the elbow (distance varied from 9.1 to 10.2 cm from the distal wrist crease).

Reference electrode (R): Placement is over the distal tendon of the flexor pollicis longus.

Ground electrode (G): Placement is on the radius, between the stimulating and recording electrodes.

Stimulation point (S): Placement is just medial to the tendon of the biceps at the elbow, slightly proximal to the antecubital crease. The shoulder is held abducted 10 degrees, with the elbow extended and the forearm supinated.

Nerve fibers tested: C7, C8, and T1 nerve roots, through the middle and lower trunks, anterior divisions, and medial and lateral cords of the brachial plexus.

Machine settings: Low frequency filter—10 Hz, High frequency filter—10 kHz.

Normal values (1) (25 subjects):

Onset latency (msec)

Mean	S.D.	Range
2.6	0.43	1.8–3.6

Upper limit of normal: 4.0 msec

Amplitude (mV)

Mean	S.D.	Range
5.6	1.16	3.8–7.5

Lower limit of normal: 2.5 mV

Helpful Hints

- The data described must be viewed with caution because the cited research has significant limitations. The data were derived from the left hands of 25 women aged 20–25 years. The authors chose the recording electrode sites to maximize the recorded amplitude, not to localize the motor point of the muscle.

- This technique may not provide for recording over the motor point. Initially positive deflections may be recorded. The waveform is characteristically bimodal and up to four phases are common.

Notes

REFERENCE

1. Craft S, Currier DP, Nelson RM: Motor conduction of the anterior interosseous nerve. *Phys Ther* 1977; 57:1143–1147.

ADDITIONAL READING/ALTERNATE TECHNIQUE

1. Baker TS, Pease WS: Anterior nerve conduction to the flexor pollicis longus (abstract). *Arch Phys Med Rehabil* 1998; 79:1149.

MEDIAN MOTOR NERVE (ANTERIOR INTEROSSEOUS BRANCH) TO THE PRONATOR QUADRATUS

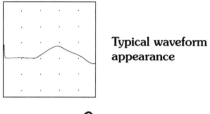

Typical waveform appearance

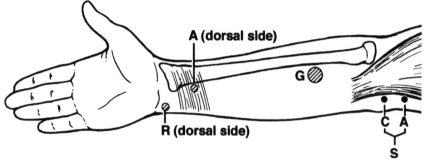

Electrode Placement

Active electrode (A): Placement is centrally over the dorsum of the forearm, 3 cm proximal to the ulnar styloid.

Reference electrode (R): Placement is over the ulnar styloid.

Ground electrode (G): Placement is over the dorsal forearm, between the stimulating and recording electrodes.

Stimulation point (S): The cathode (C) is placed over the median nerve at the elbow. The anode (A) is proximal.

Machine settings: Sensitivity for latency and duration—500 µV/division, Sensitivity for amplitude—adequate to record entire potential, Low frequency filter—20 Hz, High frequency filter—10 kHz.

Nerve fibers tested: C7, C8, and T1 nerve roots, through the middle and lower trunks, anterior divisions, and medial and lateral cords of the brachial plexus.

Normal values (1) (26 subjects) (skin temperature over the mid volar forearm 32–34 degrees Celsius):

Onset latency (msec)

	Mean	S.D.	Mean + 2 S.D.	Range
Left	3.5	0.4	4.3	2.8–4.3
Right	3.6	0.4	4.4	2.9–4.4

Onset to peak amplitude (mV)

	Mean	S.D.	Mean – 2 S.D.	Range
Left	3.1	0.8	1.5	2.0–5.5
right	3.1	0.8	1.5	2.0–5.2

Duration of negative peak (msec)

	Mean	S.D.	Mean + 2 S.D.	Range
Left	7.8	1.1	10.0	5.6–10.5
Right	7.7	1.3	10.3	5.5–10.4

Acceptable Differences

The upper limit of normal difference in latency from one side to the other is 0.4 msec.

The side to side difference in amplitude is 11.4 ± 7.7% (range 0–25%).

The side to side difference in duration is 0.8 ± 0.6 msec (range 0.0–2.4 msec).

Helpful Hints

- In obtaining the above-referenced data the subjects were comfortably seated with the elbow flexed approximately 90 degrees and with the forearm fully pronated.

- The distances between the stimulating and recording electrodes for the referenced data were equal from side to side and ranged from 17.5 to 28 cm.

- Mysiw and Colachis (1) found that this technique of surface recording produced latencies within 0.2 msec of those obtained with needle recording.

- Mysiw and Colachis (1) found that needle stimulation of just the anterior interosseous branch produced an essentially identical waveform, confirming that this technique actually records the anterior interosseous-to-pronator quadratus segment.

- A needle recording technique has also been described (2), but Shafshak and El-Hinawy (3), who compared the two techniques, thought that the surface recording technique was more sensitive to pathology than was the needle recording technique.

- There are two heads of the pronator quadratus, which may result in a bimodal evoked response. This may limit the usefulness of the duration measurement.

Notes

REFERENCES

1. Mysiw WJ, Colachis SC: Electrophysiologic study of the anterior interosseous nerve. *Am J Phys Med Rehabil* 1988; 67:50–54.
2. Nakano KK, Lundergan C, Okihiro MM: Anterior interosseous nerve syndromes: diagnostic methods and alternative treatments. *Arch Neurol* 1977; 34:477–480.
3. Shafshak TS, El-Hinaway YM: The anterior interosseous nerve latency in the diagnosis of severe carpal tunnel syndrome with unobtainable median nerve distal conduction. *Arch Phys Med Rehabil* 1995; 76:471–475.

ADDITIONAL READING/ALTERNATE TECHNIQUE

1. Baker TS, Pease WS: Anterior interosseous nerve conduction to the flexor pollicis longus (abstract). *Arch Phys Med Rehabil* 1998; 79:1149.

MEDIAN MOTOR NERVE TO THE 1ST LUMBRICAL

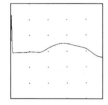

Typical waveform appearance

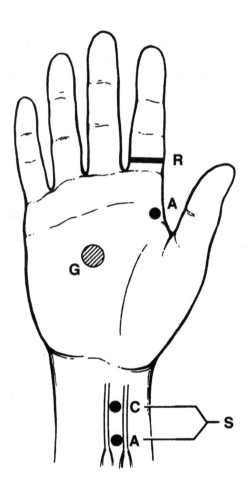

Electrode Placement

Active electrode (A): Placement is on the palm, slightly radial to the long flexor tendon of the index finger (localized by flexion of the index finger) and 1 cm proximal to the midpalmar crease.

Reference electrode (R): A ring electrode is placed at the base of the index finger.

Ground electrode (G): Placement is on the dorsum of the hand.

Stimulation point (S): The cathode (C) is placed 12 cm proximal to the active electrode, in a line following the course of the nerve from the active electrode to the interthenar crease and up the forearm. The anode (A) is proximal.

Machine settings: Sensitivity—0.5–2.0 mV/division, Low frequency filter—20 Hz, High frequency filter—10 kHz, Sweep speed—2 msec/division.

Nerve fibers tested: C8 and T1 nerve roots, through the lower trunk, anterior division, and medial cord of the brachial plexus.

Normal values (1) (44 subjects) (skin temperature over the first webspace greater than 31 degrees Celsius):

Onset latency (msec)

Mean	S.D.	Mean + 2 S.D.	Range
3.5	0.34	4.18	2.6–4.1

Amplitude (mV)

Mean	S.D.	Mean − 2 S.D.	Range
2.241	1.150	0.0	0.644–8.125

Helpful Hints

- Although probably less sensitive in detecting carpal tunnel syndrome (CTS), this test is nevertheless abnormal in some persons with CTS who have a normal latency to the abductor pollicis brevis.

- Stimulation can also be performed at the palm. If the amplitude with palm stimulation is significantly greater than with wrist stimulation, this may be a sign of neurapraxia at the wrist. The upper limit of normal increase in amplitude for palm versus wrist stimulation (mean + 2 S.D.) is 105% (mean increase 22%, range –4 to 70%). Palmar stimulation may be difficult in persons with thick skin and may also activate other nerve branches or muscles directly. The waveform shape should be the same as with wrist stimulation.

- In the cited study, the upper limit of normal latency described was nearly identical to that normally obtained on standard (8 cm) testing of the median nerve to the abductor pollicis brevis.

Notes

REFERENCE

1. Fitz WR, Mysiw WJ, Johnson EW: First lumbrical latency and amplitude: control values and findings in carpal tunnel syndrome. *Am J Phys Med Rehabil* 1990; 69:198–201.

MEDIAN MOTOR NERVE TO THE 2ND LUMBRICAL (SEE ALSO ULNAR MOTOR NERVE TO THE PALMAR INTEROSSEOUS)

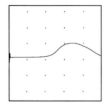

Typical waveform appearance

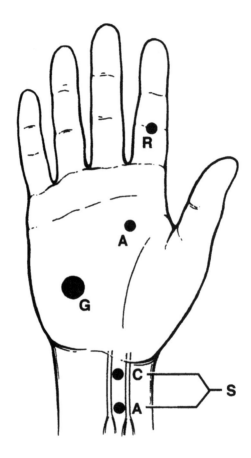

Electrode Placement

Active electrode (A): Placement is on the palm, slightly radial to the midpoint of the third metacarpal.

Reference electrode (R): Placement is over the bony prominence of the second proximal interphalangeal joint.

Ground electrode (G): Placement is between the stimulating and recording electrodes.

Stimulation point (S): Standard wrist stimulation is performed (see section on median motor nerve to the abductor pollicis brevis) The anode (A) is proximal. When comparing latencies to the second lumbrical and the interosseous muscles, identical distances are used between stimulating and recording electrodes. The cited study used distances of 8–12 cm.

Machine settings: Standard motor settings, Sensitivity—1–2 mV/division, Sweep speed—2 msec/division.

Nerve fibers tested: C8 and T1 nerve roots, through the lower trunk, anterior division, and medial cord of the brachial plexus.

Normal values (1) (51 subjects) (skin temperature over the palm greater than 31 degrees Celsius):

Onset latency (msec)

Mean	Mean + 2 S.D.	Range
3.22	3.98	2.10–4.00

Amplitude (mV)

Mean	Mean – 2 S.D.	Range
2.81	0.40	1.00–6.00

Acceptable Differences

The difference in latency between recording from the second lumbrical and the abductor pollicis brevis (APB) using the same stimulation site can be calculated. The difference is (ABP latency minus lumbrical latency) 0.0 ± 0.1 msec (range −0.2 to 0.2 msec) (2).

The difference in latency between recording from the second lumbrical (median wrist stimulation) and the interosseous muscle (ulnar wrist stimulation) using the same recording site and identical distances can be calculated. The difference (lumbrical minus interosseous) is 0.08 msec (mean + 2 S.D. = 0.40 msec, range −0.3 to 0.4 msec) (1). The upper limit of normal increase of the lumbrical versus interosseous latency is 0.4 msec. An upper limit of normal difference of 0.5 msec has also been described (3).

Helpful Hints

- Concomitant median and ulnar nerve stimulation must be avoided.

- The second lumbrical and interosseous muscles lie superimposed in this location. Stimulating the median nerve activates the lumbrical, whereas stimulating the ulnar nerve activates the interosseous muscle. Both nerve studies have approximately the same latencies and can thus be compared to detect slowing of one nerve or the other.

- Sometimes the median mixed nerve potential is seen before the desired motor response on median nerve stimulation. This potential is ignored and does not generally distort the recording from the second lumbrical.

- Anomalous innervation is common and may result in no response being seen to stimulation of one of the involved nerves.

- Conduction to the lumbricals may be relatively spared in carpal tunnel syndrome.

Notes

REFERENCES

1. Preston DC, Logigian EL: Lumbrical and interossei recording in carpal tunnel syndrome. *Muscle Nerve* 1992; 15:1253–1257.
2. Logigian EL, Busis NA, Berger AR, et al: Lumbrical sparing in carpal tunnel syndrome: anatomic, physiologic, and diagnostic implications. *Neurology* 1987; 37:1499–1505.
3. Uncini A, DiMuzio A, Awad J, et al: Sensitivity of three median-to-ulnar comparative tests in diagnosis of mild carpal tunnel syndrome. *Muscle Nerve* 1993; 16:1366–1373.

ADDITIONAL READINGS/ALTERNATE TECHNIQUES

1. Muellbacher W, Mamoli B, Zifko U, Grisold W: Lumbrical and interossei recording in carpal tunnel syndrome (letter to the editor). *Muscle Nerve* 1994; 17:359–360.
2. Seror P: The value of special motor and sensory tests for the diagnosis of benign and minor median nerve lesion at the wrist. *Am J Phys Med Rehabil* 1995; 74:124–129.
3. Sheean GL, Houser MK, Murray MF: Lumbrical-interosseous latency comparison in the diagnosis of carpal tunnel syndrome. *Electroencephalogr clin Neurophysiol* 1995; 97:285–289.

MUSCULOCUTANEOUS MOTOR NERVE TO THE BICEPS BRACHII

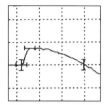

Typical waveform appearance

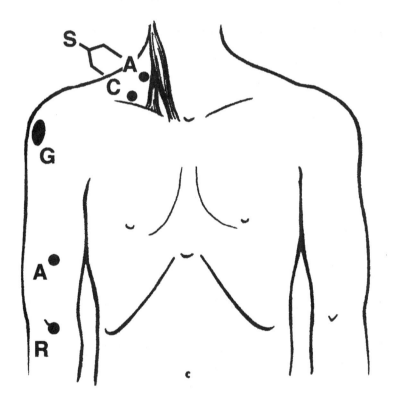

Electrode Placement

Active electrode (A): Placement is just distal to the midportion of the biceps brachii muscle.

Reference electrode (R): Placement is proximal to the antecubital fossa in the region of the junction of the muscle fibers and the biceps tendon.

Ground electrode (G): Placement is over the posterolateral shoulder.

Stimulation point (S): Erb's point—the cathode is placed slightly above the upper margin of the clavicle lateral to the clavicular head of the sternocleidomastoid muscle. The anode is superomedial.

Machine settings: Standard motor settings are used.

Nerve fibers tested: C5 and C6 nerve roots, through the upper trunk, anterior division, and lateral cord of the brachial plexus.

Normal values (1) (60 subjects) (room temperature 75–84 degrees Fahrenheit):

Onset latency (msec)

Mean	S.D.	Normal Range	Observed Range
4.5	0.6	3.3–5.7	3.2–5.9

Helpful Hints

- The active electrode is placed over the region of greatest muscle mass.
- The distance between the stimulation point and the active electrode ranges from approximately 23.5 to 29.0 cm, measured with obstetric calipers with the arm by the side.

Notes

REFERENCE

1. Kraft GH: Axillary, musculocutaneous and suprascapular nerve latency studies. *Arch Phys Med Rehabil* 1972; 53:383–387.

ADDITIONAL READING/ALTERNATE TECHNIQUE

1. Gassel MM: A test of nerve conduction to muscles of the shoulder girdle as an aid in the diagnosis of proximal neurogenic and muscular disease. *J Neurol Neurosurg Psychiatry* 1964; 27:200–205.

PHRENIC MOTOR NERVE TO THE DIAPHRAGM

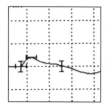

Typical waveform appearance

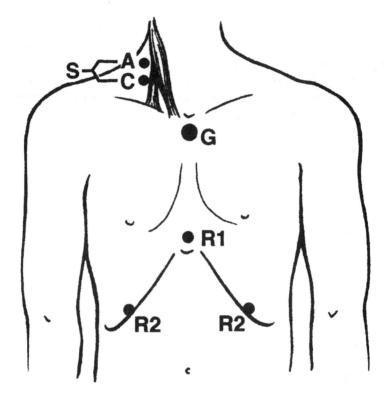

Electrode Placement

Recording electrodes (R): In the cited study, self-adhesive 2.5 cm X 2.5 cm surface electrodes were used. One electrode is placed 5 cm superior to the tip of the xiphoid process. The other electrode is placed 16 cm distally along the costal margin (usually at the 7th intercostal space) (1).

Ground electrode (G): Placement is over the upper chest (2).

Stimulation point (S): Stimulation is applied at the posterior border of the sternocleidomastoid muscle in the supraclavicular fossa, with the cathode (C) approximately 3 cm superior to the clavicle. The anode (A) is superior to the cathode. Subjects are supine, with the neck neutral or slightly extended (3). In the cited study, two supramaximal responses were obtained and the results averaged (1).

Machine settings: Low frequency filter—5 Hz, High frequency filter—5 kHz.

Nerve fibers tested: C3, C4, and C5 nerve roots.

Normal values (1) (25 subjects):

Onset latency (msec)

Mean	S.D.	Range	Suggested Normal Limit
6.54	0.77	5.5–8.4	< 8.1

Amplitude (μV)

Mean	S.D.	Range	Suggested Normal Limit
660	201	301–1198	> 300

Area (nVsec)

Mean	S.D.	Range	Suggested Normal Limit
7.28	2.09	4.0–12.8	> 4.0

Duration from onset to return to baseline (msec)

Mean	S.D.	Range	Suggested Normal Limit
19.4	2.7	13.4–24.1	< 25

Acceptable Differences

Mean side to side difference for latency is 0.34 ± 0.27 msec with a range of 0–1.2 msec. The upper limit of normal difference, based on mean + 2 S.D., is 0.88 msec or 12.6%.

Mean side to side difference for amplitude is 109 ± 94 nV with a range of 0–360 nV. The upper limit of normal difference, based on mean + 2 S.D., is 297 nVsec or 39.5%.

Mean side to side difference for area is 1.41 ± 1.26 nVsec with a range of 0.02–5.04 nVsec. The upper limit of normal difference, based on mean + 2 S.D., is 3.92 nVsec or 46.3%.

Mean side to side difference for duration is 2.26 ± 1.69 msec with a range of 0.15–6.2 msec. The upper limit of normal difference, based on mean + 2 S.D., is 5.6 msec or 28.1%.

Helpful Hints

- Increasing age is associated with increasing latency. This may need to be taken into account when studying older subjects. A larger chest circumference is associated with increased amplitude (1).

- EKG artifact may occasionally be recorded as a prolonged (> 50 msec), large amplitude response. The stimulus should be repeated until a valid response is obtained (1,3).

- Improper stimulus location may inadvertently activate the brachial plexus. This results in a volume conducted potential being recorded. The latency is shorter and there is an initial positive phase of the waveform. Brachial plexus stimulation may cause arm movement, arm paresthesias, and a short latency, low amplitude, initially positive response being recorded (1,3).

- Deep breathing should be avoided during the testing. Quiet breathing should not interfere with the results (3).

- Because of amplitude variability, it may be helpful to repeat the study several times to obtain the two highest amplitudes; these should be relatively consistent.

- Needle stimulation and more anterior surface stimulation just medial and superior to the clavicular insertion of the sternocleidomastoid muscle may also be performed (4,5).
- Side to side difference in latency with needle stimulation has been described as 0.08 ± 0.42 msec (4).

Notes

REFERENCES

1. Chen R, Collins S, Remtulla H, et al: Phrenic nerve conduction study in normal subjects. *Muscle Nerve* 1995; 18:330–335.
2. Markand ON, Kinaid JC, Pourmand RA: et al: Electrophysiologic evaluation of diaphragm by transcutaneous phrenic nerve stimulation. *Neurology* 1984; 34:604–614.
3. Bolton CF: AAEM minimonograph #40: clinical neurophysiology of the respiratory system. *Muscle Nerve* 1993; 16:809–818.
4. MacLean IC, Mattioni TA: Phrenic nerve conduction studies: a new technique and its application in quadriplegic patients. *Arch Phys Med Rehabil* 1981; 62:70–73.
5. Ma DM, Liveson JA: *Nerve conduction handbook*. Philadelphia: FA Davis, 1983.

ADDITIONAL READINGS/ALTERNATE TECHNIQUES

1. Newsom Davis J: Phrenic nerve conduction in man. *J Neurol Neurosurg Psychiatry* 1967; 30:420–426.
2. Swenson MR, Rubenstein RS: Phrenic nerve conduction studies. *Muscle Nerve* 1992; 15:597–603.
3. Russell RI, Helps BA, Elliot MJ, Helms PJ: Phrenic nerve stimulation at the bedside in children: equipment and validation. *Eur Respir J* 1993; 6:1332–1335.

RADIAL NERVE

RADIAL (POSTERIOR INTEROSSEOUS) MOTOR NERVE TO THE EXTENSOR CARPI ULNARIS AND BRACHIORADIALIS

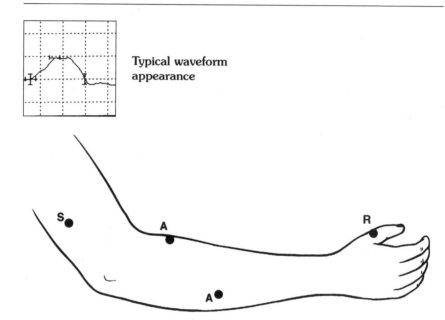

Typical waveform appearance

Electrode Placement

Active electrode (A): For the brachioradialis, placement is on the belly of the muscle, 3 cm distal to the elbow. For the extensor carpi ulnaris, placement is at the mid-forearm (equal distance between the lateral epicondyle and the ulnar styloid process), close to the "ulnar crease."

Reference electrode (R): Placement is on the thumb.

Stimulation point (S): The cathode is a monopolar needle electrode inserted 5–6 cm proximal to the lateral epicondyle on the lateral upper arm. The anode is a subcutaneous needle electrode located 2 cm proximally.

Machine settings: Sensitivity—2–5 mV/division, Sweep speed—3 msec/division.

Nerve fibers tested: Extensor carpi ulnaris: C6, C7, and C8 nerve roots, through the upper, middle, and lower trunks, posterior divisions, and posterior cord of the brachial plexus, then through the radial nerve and the posterior interosseous branch of the radial nerve. Brachioradialis: C5 and C6 nerve roots, through the upper trunk, posterior division, and posterior cord of the brachial plexus.

Normal values (1) (40 subjects—data for both sides combined) (skin temperature over palm and forearm greater than 31 degrees Celsius):

Onset latency (msec)

Brachioradialis

Mean	S.D.	Range	Upper Limit of Normal
2.66	0.32	1.8–3.5	3.3

Extensor carpi ulnaris

Mean	S.D.	Range	Upper Limit of Normal
4.00	0.35	3.1–5.2	4.17

Acceptable Differences

The upper limit of normal difference in latency between the extensor carpi ulnaris versus the brachialis is 1.8 msec (mean 1.34 ± 0.23, range 0.8–2.0).

The upper limit of normal side to side difference in latency is 0.4 msec (mean 0.19 ± 0.06, range 0.0–0.4).

Helpful Hints

- The nerve branch to the brachioradialis does not pass through the "radial tunnel," whereas the branch to the extensor carpi ulnaris does.
- The author states that needle stimulation is preferable to surface stimulation because surface stimulation at this point often requires painful high intensity stimulation and often causes electrical artifacts. Needle stimulation also localizes the stimulation site more precisely.

Notes

REFERENCE

1. Seror P: Posterior interosseous nerve conduction: a new method of evaluation. *Am J Phys Med Rehabil* 1996; 75:35–39.

RADIAL MOTOR NERVE TO THE EXTENSOR DIGITORUM COMMUNIS

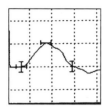

Typical waveform appearance

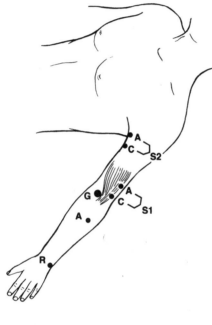

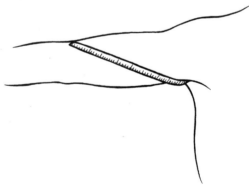

Electrode Placement

Upper Extremity/Cervical Plexus/Brachial Plexus Motor Studies

Active electrode (A): Placement is 8 cm distal to stimulation point 1 over the extensor digitorum communis. This site is localized in the reference (1) by grasping the radius and ulna of the subject's pronated forearm with the thumb and middle finger at the junction of the upper third and middle third of the forearm. The index finger is placed halfway between these two points to identify the extensor digitorum communis.

Reference electrode (R): Placement is over the ulnar styloid process.

Ground electrode (G): Placement is between the stimulating and recording electrodes.

Stimulation point 1 (S1): The cathode (C) is placed in the antecubital fossa just lateral to the biceps tendon as the tendon crosses the flexor crease. The anode (A) is proximal. The subject is supine, and the arm is supported and abducted 40–45 degrees.

Stimulation point 2 (S2): The cathode (C) is placed in the axilla between the coracobrachialis and the long head of the triceps. The anode (A) is proximal.

Machine settings: Sensitivity—5 mV/division, Low frequency filter—5 Hz, High frequency filter—10 kHz, Sweep speed—5 msec/division.

Nerve fibers tested: C7 and C8 nerve roots, through the middle and lower trunks, posterior divisions, and posterior cord of the brachial plexus, then through the radial nerve and posterior interosseous branch of the radial nerve.

Normal values For the right side (left side results were similar) (1) (30 subjects) (skin temperature over the forearm greater than or equal to 34 degrees Celsius):

Onset latency (msec)

Mean	S.D.	Mean + 2 S.D.	Normal Range
2.6	0.44	3.48	< 3.5

Amplitude (mV)

Mean	S.D.	Mean − 2 S.D.	Normal Range
11.31	3.5	4.31	> 4.4

Nerve conduction velocity between S1 and S2 (m/sec)—the distance is measured from the axilla, across the biceps anteriorly (see figure), to the elbow stimulation site (caliper measurement should give a similar distance measurement) (2).

Mean	S.D.	Mean − 2 S.D.	Normal Range
68	7.0	54	> 51

Helpful Hints

- With proximal stimulation there can be a problem of recording volume conducted potentials from other muscles. Therefore, only the minimal stimulus intensity necessary to produce a waveform similar in appearance to that on distal stimulation is recommended.
- Rotation of the anode may be necessary to obtain an optimal recording.

Notes

REFERENCES

1. Young AW, Redmond MD, Hemler DE, Belandres PV: Radial motor nerve conduction studies. *Arch Phys Med Rehabil* 1990; 71:399–402.
2. Kalantri A, Visser BD, Dumitru D, Grant AE: Axilla to elbow radial nerve conduction. *Muscle Nerve* 1988; 11:133–135

RADIAL MOTOR NERVE TO THE EXTENSOR INDICIS PROPRIUS

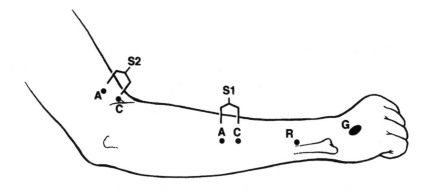

Electrode Placement

Recording electrodes (R): A concentric needle electrode is placed into the extensor indicis proprius on the dorsal forearm (1). The muscle is slightly radial to the ulna and extensor carpi ulnaris tendon, approximately 4 cm proximal to the ulnar styloid process, and approximately one half inch deep. Monopolar needle electrode recording has also been described with a surface reference electrode placed on the 5th digit (2,3).

Ground electrode (G): Placement is over the dorsum of the hand or between the stimulating and recording electrodes.

Stimulation point 1 (S1): The cathode (C) is placed 3–4 cm proximal to the needle insertion site between the extensor carpi ulnaris and the extensor digiti minimi. The anode (A) is proximal (3).

Stimulation point (S2): The cathode (C) is placed 5–6 cm proximal to the lateral epicondyle in the groove between the brachialis and brachioradialis muscles. The anode (A) is proximal.

Stimulation point 3: The stimulating electrodes are placed at Erb's point.

Nerve fibers tested: C7 and C8 nerve roots, through the middle and lower trunks, posterior divisions, and posterior cord of the brachial plexus, then through the radial nerve and the posterior interosseous branch of the radial nerve.

Machine settings: Standard motor settings are used.

Normal values:

Onset latency over 2.8–6.6 cm distance (msec) (29 subjects, monopolar needle recording) (2)

Mean	S.D.	Range
1.69	0.29	1.0–2.0

Nerve conduction velocity *(the distal segment was measured with a tape measure, the proximal segment with obstetric calipers; arm abducted 10 degrees, elbow flexed 10–15 degrees, forearm pronated, head rotated away from side being tested)*

S1–S2 nerve conduction velocity (m/sec) (49 subjects) (1)

Mean	S.D.	Range
61.6	5.9	48–75

S2–S3 nerve conduction velocity (m/sec) (49 subjects) (1)

Mean	S.D.	Range
72.0	6.3	56–93

Acceptable Difference

If the proximal velocity is less than 60 m/sec or if the distal velocity is more than 6 m/sec faster than the proximal velocity, a disturbance of conduction in the proximal segment may be suspected (1).

Helpful Hints

- Surface or needle recording has been described. It is important that the shape of the waveform be similar with proximal and distal stimulation (2,3).

- The site of needle insertion can usually be localized by first having the subject flex and extend the index finger while palpating the muscle. The needle is inserted and proper placement is confirmed by free run electromyogram.

- The preceding nerve conduction velocity results were obtained with a concentric needle recording, but the same author also described monopolar needle recording from the same muscle (3). He found that this did not consistently provide a negative takeoff with Erb's point stimulation. It seems reasonable to substitute a monopolar needle as long as proper care is given to recording an accurate onset of the waveform.

- An armboard may help to stabilize the forearm and prevent needle movement (1).
- A needle stimulation technique has also been described and recommended as more accurate than surface stimulation (4,5).
- Axillary stimulation can also be performed. In a study that utilized needle stimulation and recording, the distal latency was 2.4 ± 0.5 msec, axilla to above elbow nerve conduction velocity was 69 ± 5.6m/sec, and above elbow to forearm nerve conduction velocity was 62 ± 5.1 m/sec (4).

Notes

REFERENCES

1. Jebsen RH: Motor conduction velocity in proximal and distal segments of the radial nerve. *Arch Phys Med Rehabil* 1966; 47:597–602.
2. Ma DM, Liveson JA: *Nerve conduction handbook.* Philadelphia: FA Davis, 1983.
3. Jebsen RH: Motor conduction velocity of distal radial nerve. *Arch Phys Med Rehabil* 1966; 47:12–16.
4. Trojaborg W, Sindrup EH: Motor and sensory conduction in different segments of the radial nerve in normal subjects. *J Neurol Neurosurg Psychiatry* 1969; 32:354–359.
5. Falck B, Hurme M: Conduction velocity of the posterior interosseous nerve across the arcade of Frohse. *Electromyogr clin Neurophysiol* 1983; 23:567–576.

ADDITIONAL READING/ALTERNATE TECHNIQUE

1. Gassel MM, Diamantopoulos E: Pattern of conduction times in the distribution of the radial nerve. *Neurology* 1964; 14:222–231.

SUPRASCAPULAR MOTOR NERVE TO THE SUPRASPINATUS AND INFRASPINATUS

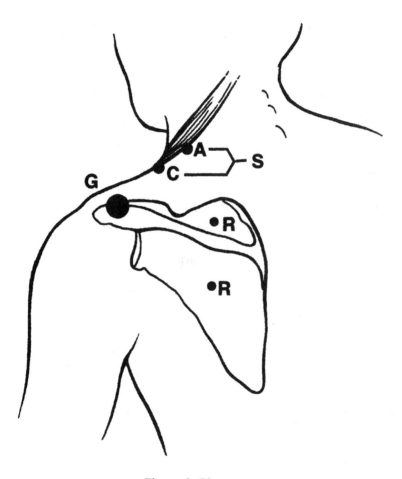

Electrode Placement

Recording electrodes (R): A coaxial needle is inserted into the deep fibers of the muscles. For the supraspinatus the midpoint of the spine of the scapula is identified by palpation. The needle is inserted medial to this point just above the spine in a downward and forward direction until the scapula is touched. The needle is then withdrawn several millimeters. Placement is confirmed by asking the subject to abduct the arm. For the infraspinatus the needle is inserted several centimeters below the scapular spine and several centimeters lateral to the medial border of the scapula. The needle is again inserted to the bone and withdrawn a few millimeters. Needle placement is confirmed with active external rotation.

Ground electrode (G): Placement is over the posterolateral shoulder.

Stimulation point (S): Erb's point—the cathode (C) is placed slightly above the upper margin of the clavicle lateral to the clavicular head of the sternocleidomastoid muscle. The anode (A) is superomedial.

Machine settings: Standard motor settings are used.

Nerve fibers tested: C5 and C6 nerve roots, through the upper trunk of the brachial plexus.

Normal values (1) (room temperature 75–84 degrees Fahrenheit):

Onset latency to the supraspinatus (msec—62 subjects)

Mean	S.D.	Normal Range	Observed Range
2.7	0.5	1.7–3.7	1.9–3.8

Onset latency to the infraspinatus (msec—60 subjects)

Mean	S.D.	Normal Range	Observed Range
3.3	0.5	2.4–4.2	2.4–4.4

Helpful Hints

- The technique is performed with the subject sitting with his arms by his sides.

- Large amplitude polyphasic potentials are expected with this needle recording technique.

- Too shallow a needle placement may result in a suboptimal recording, and if very shallow, the needle may even be in the trapezius muscle.

- If monopolar needles are used for the recording, a reference needle can be inserted approximately 2 cm from the active electrode (2).

- The distance between the stimulation point and the needle electrode ranges from approximately 7.4 to 12.0 cm for the supraspinatus recording and from approximately 10.6 to 15.0 cm for the infraspinatus recording, measured with obstetric calipers.

- A surface recording technique has also been described (3), studying 10 subjects with stimulation at Erb's point and recording over the supraspinatus and infraspinatus. The mean amplitude over the supraspinatus was 10.2 ± 2.78 mV (range 6.4–16.9). The mean amplitude over the infraspinatus was 11.2 ± 3.58 mV (range 5.0–17.6) The mean side to side difference in amplitude was 14.6%, with the largest side to side difference being a 40% drop.

- In another study, latencies were similar to those described, but side to side differences were also investigated. In 90% of normal subjects the side to side latency difference was less than 0.4 msec. In 50% of patients with suprascapular neuropathy the side to side difference was greater than 0.4 msec. The authors concluded that a side to side difference of more than 0.4 msec is a strong predictor of suprascapular neuropathy (4).

Notes

REFERENCES

1. Kraft GH: Axillary, musculocutaneous and suprascapular nerve latency studies. *Arch Phys Med Rehabil* 1972; 53:383–387.
2. Ma DM, Liveson JA: *Nerve conduction handbook.* Philadelphia: FA Davis, 1983.
3. Clark JD, King RB, Ashman E: Detection of suprascapular nerve lesions using surface recording electrodes. Presented at the AAEM annual meeting, Sept. 20, 1997.
4. Edgar TS, Lotz BP: A nerve conduction technique for the evaluation of suprascapular neuropathies. Presented at the AAEM annual meeting, Orlando, Florida, October 16, 1998.

ADDITIONAL READING/ALTERNATE TECHNIQUE

1. Gassel MM: A test of nerve conduction to muscles of the shoulder girdle as an aid in the diagnosis of proximal neurogenic and muscular disease. *J Neurol Neurosurg Psychiatry* 1964; 27:200–205.

THORACODORSAL MOTOR NERVE TO THE LATISSIMUS DORSI

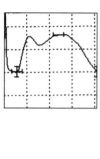

Typical waveform appearance

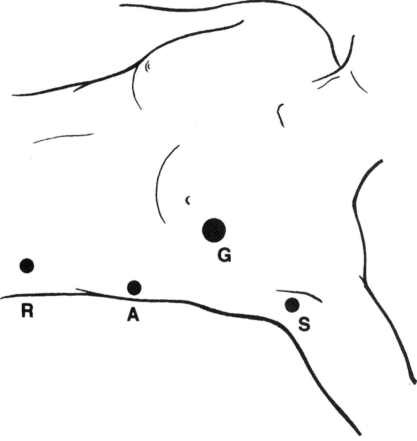

Electrode Placement

Active electrode (A): Placement is on the posterior axillary line at the level of the inferior pole of the scapula.

Reference electrode (R): Placement is on the ipsilateral flank.

Ground electrode (G): Placement is on the ipsilateral lateral chest wall.

Stimulation point (S): The cathode is placed in the axilla with the anode proximal. The subject is supine, with the arm abducted to 90 degrees.

Machine settings: Sensitivity—2 mV/division, Low frequency filter—2 Hz, High frequency filter—10 kHz, Sweep speed—1 msec/division, Pulse duration—0.2 msec.

Nerve fibers tested: C6, C7, and C8 nerve roots, through the upper, middle, and lower trunks, posterior divisions, and posterior cord of the brachial plexus.

Normal values (1) (30 subjects—right side data):

Onset latency (msec)

Mean	S.D.	Mean + 2 S.D.	Range
1.9	0.4	2.7	1.2–2.7

Amplitude (mV)

Mean	S.D.	Range
4.1	1.8	1.4–10.2

Acceptable Difference

The upper limit of normal decrease in amplitude from one side to the other is 50%.

Helpful Hints

- The latissimus dorsi can be localized by asking the subject to depress and internally rotate the shoulder.
- The distance between the stimulation point and the active electrode ranged from 5 to 12 cm, measured with a tape measure with the arm abducted 90 degrees.
- In obese subjects it may be helpful to press the stimulator deeper into the axilla toward the lateral margin of the scapula to obtain a response.
- In the cited study, Erb's point stimulation was performed to calculate a conduction velocity across the axillary segment. This measure was not deemed to be reliable.

Notes

REFERENCE

1. Wu PBJ, Robinson T, Kingery WS, Date ES: Thoracodorsal nerve conduction study. *Am J Phys Med Rehabil* 1998; 77:296–298.

ADDITIONAL READING/ALTERNATE TECHNIQUE

1. Lo Monaco M, Di Pasqua PG, Tonali P: Conduction studies along the accessory, long thoracic, dorsal scapular, and thoracodorsal nerves. *Acta Neurol Scand* 1983; 68:171–176.

ULNAR NERVE

ULNAR MOTOR NERVE TO THE ABDUCTOR DIGITI MINIMI

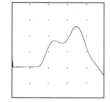
Typical waveform appearance

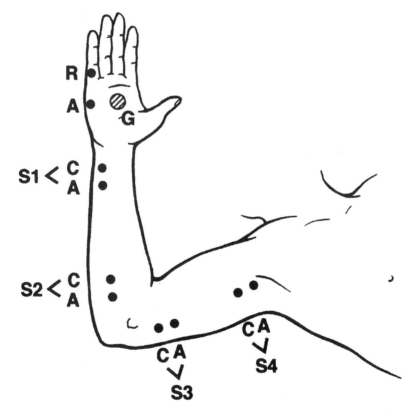
Electrode Placement

For this study the arm is positioned in a 45 degree abducted and externally rotated posture. The elbow is flexed to 90 degrees (right angle) and the forearm is in neutral position (thumb pointing toward the ear).

Active electrode (A): Placement is on the ulnar surface of the hypothenar eminence, halfway between the level of the pisiform bone and the 5th metacarpophalangeal joint.

Reference electrode (R): Placement is slightly distal to the 5th metacarpophalangeal joint.

Ground electrode (G): Placement is on the dorsum of the hand. If stimulus artifact interferes with the recording, the ground may be placed near the active electrode, between this electrode and the cathode.

Stimulation point 1 (S1): The cathode (C) is placed 8 cm proximal to the active electrode, in a line measured slightly radial to the tendon of the flexor carpi ulnaris. The anode (A) is proximal.

Stimulation point 2 (S2): The cathode (C) is placed approximately 4 cm distal to the medial epicondyle. The anode (A) is proximal.

Stimulation point 3 (S3): The cathode (C) is placed approximately 10 cm proximal to stimulation point 2, measured in a curve behind the medial epicondyle to a point slightly volar to the triceps. The anode (A) is proximal.

Stimulation point 4 (S4): The cathode (C) is placed in the axilla approximately 10 cm proximal to stimulation point 3. The anode (A) is proximal.

F-wave stimulation: The cathode (C) is positioned as for stimulation point 1, but with the anode distal.

Machine settings: Sensitivity—5 mV/division, Low frequency filter—2–3 Hz, High frequency filter—10 kHz, Sweep speed—2 msec/division.

Nerve fibers tested: C8 and T1 nerve roots, through the lower trunk, anterior division, and medial cord of the brachial plexus.

Normal values (1) (248 subjects) (skin temperature over the dorsum of the hand greater than or equal to 32 degrees Celsius):

Onset latency (msec)

Mean	S.D.	Mean + 2 S.D.	97th Percentile
3.0	0.3	3.6	3.7

Amplitude (mV)

Mean	S.D.	Mean − 2 S.D.	3rd Percentile
11.6	2.1	7.4	7.9

Area of negative phase (µVsec)

Mean	S.D.	Mean − 2 S.D.	3rd Percentile
35.9	7.1	21.7	23.9

Duration of negative phase (msec)

Mean	S.D.	Mean + 2 S.D.	97th Percentile
6.0	0.9	7.8	7.7

Nerve conduction velocity (m/sec)

	Mean	S.D.	Mean − 2 S.D.	3rd Percentile
S1–S2	61	5	51	52
S2–S3	61	9	43	43
S3–S4	61	7	47	50

F-wave latencies (msec) (2) (193 subjects)—shortest of 10 stimuli

Age 19–49

Height in cm (in.)	Mean	S.D.	Mean + 2 S.D.	97th Percentile
< 160 (5'3")	23.5	1.3	26.1	
160–179 (5'3"–5'10")	26.2	2.0	30.2	
≥ 180 (5'11")	29.2	1.8	32.8	

Age 50–79

Height in cm (in.)	Mean	S.D.	Mean + 2 S.D.	97th Percentile
< 160 (5'3")	25.0	1.9	28.8	
160–179 (5'3"–5'10")	28.1	1.4	30.9	
≥ 180 (5'11")	30.4	1.7	33.8	
All subjects	26.5	2.5	31.5	31.1

Acceptable Differences

The upper limit of normal increase in latency from one side to the other is 0.6 msec.

The upper limit of normal decrease in amplitude from one side to the other is 25%.

The upper limit of normal decrease in nerve conduction velocity from the S1–S2 to S2–S3 segment is 15 m/sec.

The upper limit of normal decrease in S1–S2 nerve conduction velocity from one side to the other is 9 m/sec.

The upper limit of normal decrease in S2–S3 nerve conduction velocity from one side to the other is 17 m/sec.

The upper limit of normal decrease in S3–S4 nerve conduction velocity from one side to the other is 16 m/sec.

The upper limit of normal decrease in amplitude from S1 to S2 stimulation is 35%.

The upper limit of normal decrease in amplitude from S2 to S3 stimulation is 16%.

The upper limit of normal decrease in amplitude from S3 to S4 stimulation is 21%.

The upper limit of normal side to side difference in the shortest F-wave latency is 2.5 msec.

Helpful Hints

- More proximal stimulation can also be performed at Erb's point in the supraclavicular fossa. This can allow determination of waveform changes across a more proximal segment of the nerve and calculation of more proximal nerve conduction velocity. When calculating the conduction velocity of the Erb's point-to-axilla segment, obstetric calipers are used to measure the distance.

- Anomalous innervation due to a Martin-Gruber (median to ulnar) anastomosis in the forearm is common, although it is much less commonly clinically significant during electrodiagnostic studies. It may result in a larger than expected ulnar compound motor action potential amplitude with wrist as opposed to more proximal nerve stimulation, but is usually not a confounding factor in ulnar nerve studies. If suspected, it can be investigated as described for the median motor nerve study to the abductor pollicis brevis.

- The ulnar nerve motor response to the abductor digiti minimi may be normal in Guyon's canal entrapment neuropathy at the wrist, as this muscle is usually innervated by the superficial palmar branch of the ulnar nerve. If such a compression is suspected, the motor responses to the first dorsal interosseous or palmar interosseous muscles should be studied.

- Recording a response with S2 stimulation may at times be difficult, especially in obese or muscular individuals. It may be necessary to move the cathode around to find the optimal stimulation site. Increased stimulus intensity or duration or needle stimulation may occasionally be needed. When moving from the S2 to S3 site, the intensity should again be lowered because the nerve is much easier to stimulate at this site.

  Occasionally an optimal amplitude cannot be obtained at S2 stimulation. This may give the false impression of a conduction block in the forearm. If S3 stimulation provides a normal amplitude, such a conduction block is not present. If the supramaximal S3 amplitude is not normal, a conduction block may be present. When a maximal S2 amplitude cannot be obtained, the conduction velocity calculation may be erroneous.

- An "inching technique" called short segment incremental stimulation (SSIS) can be performed to localize the site of an ulnar neuropathy at the elbow. First the nerve's course is mapped out with subthreshold stimuli by moving the stimulator perpendicular to the nerve's course until the maximal M-wave amplitude for a given subthreshold intensity is obtained. This point is marked with a dot. This process is repeated along the length of the nerve, and the dots are joined to outline the course of the nerve. Then supramaximal stimulation is performed in 1 cm increments along the length of the nerve taking care not to apply excessively supramaximal stimulation. The upper limit of normal segmental latency change is 0.4 msec. Abrupt changes in waveform shape or amplitude may be signs of local conduction block (3). The upper limit of normal segmental latency change recorded in 2 cm increments (elbow fixed at 90 degrees of flexion) has also been studied and described as 0.60–0.63 msec (4).

- Ulnar neuropathy at the elbow may be due to compression at any of three sites: the retroepicondylar groove, the humeroulnar aponeurotic arcade, and the deep forearm aponeurosis at the point of exit from under the flexor carpi ulnaris (Pridgeon's point). If possible, it is advisable to try to localize an ulnar neuropathy to one or more of these sites through incremental stimulation (3).

- The terms *cubital tunnel syndrome* and *tardy ulnar palsy* are poorly defined, are often if not usually misapplied, and should be discarded. The term *ulnar neuropathy at the elbow* (UNE) should be used instead.

Notes

REFERENCES

1. Buschbacher RM: Ulnar nerve motor conduction to the abductor digiti minimi. *Am J Phys Med Rehabil* 1999; 78:S9–S14.
2. Buschbacher RM: Ulnar nerve F-waves. *Am J Phys Med Rehabil* 1999; 78:S38–S42.
3. Campbell WW, Pridgeon RM, Sahni KS: Short segment incremental studies in the evaluation of ulnar neuropathy at the elbow. *Muscle Nerve* 1992; 15:1050–1054.
4. Kanakamedala RV, Simons DG, Porter RW, Zucker RS: Ulnar nerve entrapment at the elbow localized by short segment stimulation. *Arch Phys Med Rehabil* 1988; 69:959–963.

ADDITIONAL READINGS/ALTERNATE TECHNIQUES

1. Melvin JL, Harris DH, Johnson EW: Sensory and motor conduction velocities in the ulnar and median nerves. *Arch Phys Med Rehabil* 1966; 47:511–519.
2. Checkles NS, Russakov AD, Piero DL: Ulnar nerve conduction velocity: effect of elbow position on measurement. *Arch Phys Med Rehabil* 1971; 52:362–365.
3. Melvin JL, Schuchmann JA, Lanese RR: Diagnostic specificity of motor and sensory nerve conduction variables in the carpal tunnel syndrome. *Arch Phys Med Rehabil* 1973; 54:69–74.
4. Kincaid JC, Phillips LH, Daube JR: The evaluation of suspected ulnar neuropathy at the elbow: normal conduction study values. *Arch Neurol* 1986; 43:44–47.

5. Perez MC, Sosa A, Acevedo CEL: Nerve conduction velocities: normal values for median and ulnar nerves. *Bol Asoc Med P Rico* 1986; 78:191–196.
6. Falco FJE, Hennessey WJ, Braddom RL, Goldberg G: Standardized nerve conduction studies in the upper limb of the healthy elderly. *Am J Phys Med Rehabil* 1992; 71:263–271.
7. Hennessey WJ, Falco FJE, Braddom RL: Median and ulnar nerve conduction studies: normative data for young adults. *Arch Phys Med Rehabil* 1994; 75:259–264.
8. Campbell WW: The value of inching techniques in the diagnosis of focal nerve lesions. *Muscle Nerve* 1998; 21:1154–1561.

ULNAR MOTOR NERVE TO THE PALMAR INTEROSSEOUS (SEE ALSO MEDIAN MOTOR NERVE TO THE 2ND LUMBRICAL)

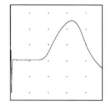

Typical waveform appearance

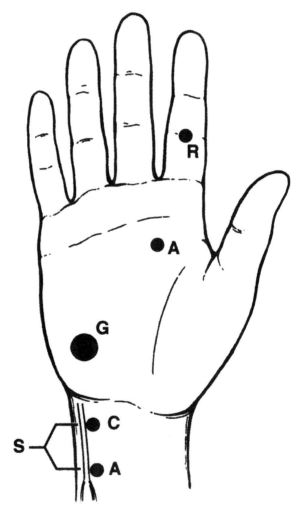

Electrode Placement

Active electrode (A): Placement is on the palm, slightly radial to the midpoint of the third metacarpal.

Reference electrode (R): Placement is over the bony prominence of the second proximal interphalangeal joint.

Ground electrode (G): Placement is between the stimulating and recording electrodes.

Stimulation point (S): Standard wrist stimulation is performed (see section on ulnar motor nerve to the abductor digiti minimi) The anode (A) is proximal to the cathode (C). When comparing latencies to the second lumbrical and the interosseous muscles, identical distances are used between the stimulating and recording electrodes.

The cited study used distances of 8–12 cm.

Machine settings: Standard motor settings are used, Sensitivity—1–2 mV/division, Sweep speed—2 msec/division.

Nerve fibers tested: C8 and T1 nerve roots, through the lower trunk, anterior division, and medial cord of the brachial plexus.

Normal values (1) (51 subjects) (skin temperature over the palm greater than 31 degrees Celsius):

Onset latency (msec)

Mean	Mean + 2 S.D.	Range
3.15	3.91	2.00–4.00

Amplitude (mV)

Mean	Mean – 2 S.D.	Range
6.55	1.67	2.60–13.60

Acceptable Difference

The difference in latency between recording from the second lumbrical (median wrist stimulation) and the interosseous muscles (ulnar wrist stimulation) using the same recording site and identical distances can be calculated. The difference (lumbrical minus interosseous) is 0.08 msec (mean + 2 S.D. = 0.40 msec, range –0.3 to 0.4) (1).

Helpful Hints

- Concomitant median and ulnar nerve stimulation must be avoided.
- The second lumbrical and interosseous muscles lie superimposed in this location. Stimulating the median nerve activates the lumbrical, whereas stimulating the ulnar nerve activates the interosseous muscle. Both nerve studies have approximately the same latencies and can thus be compared to detect slowing of one nerve or the other.
- Anomalous innervation is common and may result in no response being seen to stimulation of one of the involved nerves.

Notes

REFERENCE

1. Preston DC, Logigian EL: Lumbrical and interossei recording in carpal tunnel syndrome. *Muscle Nerve* 1992; 15:1253–1257.

ADDITIONAL READING/ALTERNATE TECHNIQUE

1. Kothari MJ, Preston DC, Logigian EL: Lumbrical-interossei motor studies localize ulnar neuropathy at the wrist. *Muscle Nerve* 1996; 19:170–174.

ULNAR MOTOR NERVE TO THE 1ST DORSAL INTEROSSEOUS

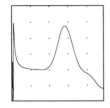

Typical waveform appearance

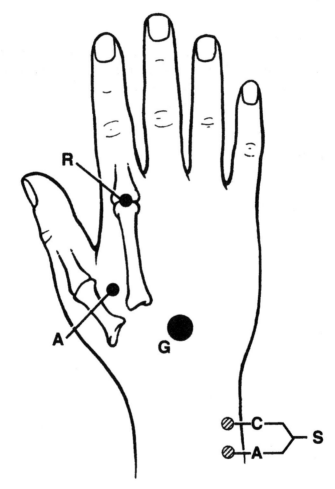

Electrode Placement

Active electrode (A): Placement is on the dorsum of the hand, over the center of the fleshy belly of the first dorsal interosseous muscle.

Reference electrode (R): Placement is over the second metacarpophalangeal joint.

Ground electrode (G): Placement is between the stimulating and recording electrodes.

Stimulation point (S): For the cited study, the cathode (C) was placed over the ulnar nerve at the proximal wrist crease (usually 5.5–6.5 cm proximal to the active electrode placement for the abductor digiti minimi—see section on ulnar motor nerve to the abductor digiti minimi) (1,2). The anode (A) is proximal.

Nerve fibers tested: C8 and T1 nerve roots, through the lower trunk, anterior division, and medial cord of the brachial plexus and the deep palmar branch of the ulnar nerve.

Machine settings: Standard motor settings are used.

Normal values: (1) (188 subjects) (Skin temperature over first dorsal interosseous muscle greater than 32 degrees Celsius):

Onset latency (msec)

Age	Mean	Range
< 20	3.3	2.7–4.2
20–29	3.4	2.6–4.1
30–39	3.3	2.5–4.4
40–49	3.2	2.3–4.2
50–59	3.4	2.6–4.4
60–69	3.6	3.0–4.5
> 70	3.6	3.0–4.2

While the mean latencies increased with advancing age, the maximum observed latencies did not correlate with age. The authors of the cited study use an upper limit of normal latency of 4.5 msec for a broad age range.

Upper Extremity/Cervical Plexus/Brachial Plexus Motor Studies

Amplitude (mV)

Age	Mean	Range
< 20	15	8–23
20–29	14	8–22
30–39	15	6–24
40–49	13	6–22
50–59	13	6–20
60–69	12	7–20
> 70	12	8–15

Although the mean amplitude decreased with advancing age, the lower limit of amplitude did not. The authors of the cited study use a lower limit of normal amplitude of 6 mV.

Acceptable Differences

The upper limit of normal increase in latency from one side to the other is 1.3 msec (mean 0.2 msec, range 0.0–1.3).

The upper limit of normal difference in latency between recording from the 1st dorsal interosseous versus the abductor digiti minimi (1st dorsal interosseous latency greater than abductor digiti minimi latency) with the same stimulation point is 2.0 msec (mean 0.9 msec, range 0.2–2.0).

Helpful Hints

- If information about more proximal ulnar nerve conduction velocity is desired, it should be obtained by studying the nerve to the abductor digiti minimi. Stimulation at the more proximal sites often activates both the median and the ulnar nerves, which causes volume conduction artifact to be recorded when studying the 1st dorsal interosseous muscle.
- Anomalous innervation of the 1st dorsal interosseous muscle is sometimes present (3).

- The waveform with this recording frequently has an initially positive deflection. Latency is measured to the initial deflection from baseline.
- In 80% of subjects the amplitude to the 1st dorsal interosseous is greater than that recorded from the abductor digiti minimi.

Notes

REFERENCES

1. Olney RK, Wilbourn AJ: Ulnar nerve conduction study of the first dorsal interosseous muscle. *Arch Phys Med Rehabil* 1985; 66:16–18.
2. Olney RK, Hanson M: AAEE case report #15: ulnar neuropathy at or distal to the wrist. *Muscle Nerve* 1988; 11:828–832.

3. Hollinshead WH, Jenkins DB: *Functional anatomy of the limbs and back*, 5th ed. Philadelphia: WB Saunders, 1981, p. 179.

ADDITIONAL READINGS/ALTERNATE TECHNIQUES

1. Bhala RP, Goodgold J: Motor conduction in the deep palmar branch of the ulnar nerve. *Arch Phys Med Rehabil* 1968; 49:460–466.
2. Seror P: Comparison of the distal motor latency of the first dorsal interosseous with abductor pollicis brevis. *Electromyogr clin Neurophysiol* 1988; 28:341–345.

Chapter 2

Upper Extremity
Sensory and
Mixed Nerves

LATERAL ANTEBRACHIAL CUTANEOUS SENSORY NERVE STUDY

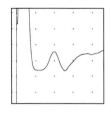

Typical waveform appearance

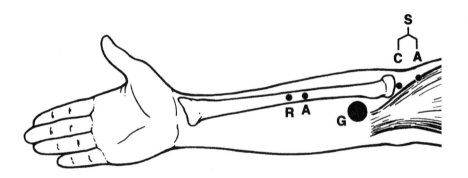

Electrode Placement

Izzo et al. Technique

Recording electrodes: A 3 cm bar electrode is placed on the radial aspect of the volar forearm, with the active electrode (A) 14 cm distal to the stimulation point. The reference electrode (R) is distal.

Ground electrode (G): Placement is between the stimulating and recording electrodes.

Stimulation point (S): The cathode (C) is placed at the lateral border of the biceps tendon in the cubital fossa. The anode (A) is proximal.

Machine settings: Standard sensory settings are used, with a sensitivity adequate to record 1 μV potentials. Averaging of eight responses was performed to derive the data presented.

Nerve fibers tested: C5 and C6 nerve roots, through the upper trunk, anterior division, and lateral cord of the brachial plexus. This is the continuation of the musculocutaneous nerve.

Normal values: (1) (157 subjects—absent in 2) (skin temperature over the forearm 30–33 degrees Celsius):

Peak latency (msec)

Mean	S.D.	Mean + 2 S.D.	Range
2.8	0.2	3.2	2.2–3.3

Peak to peak amplitude (μV)

Mean	S.D.	Range	Lower Limit of Normal (5th Percentile)
18.9	9.9	5.0–70.0	8

Kimura and Ayyar Technique

Recording electrodes: The active electrode (A) is placed 10 cm distal to the cathode along a straight line connecting the cathode to the radial artery at the wrist. The reference electrode (R) is placed approximately 3 cm distally.

Ground electrode (G): Placement is between the stimulating and recording electrodes.

Stimulation point (S): The cathode (C) is placed lateral to the biceps tendon at the level of the elbow crease. The anode (A) is proximal.

Machine settings: Low frequency filter—32 Hz, High frequency filter—3.2 Hz.

Normal values (2) (40 subjects):

Onset latency (msec)

Mean	S.D.	Mean + 2 S.D.	Range
1.57	0.19	1.95	1.2–2.0

Side to side difference: 0.05 ± 0.05 (range –0 to 0.2)

Peak latency (msec)

Mean	S.D.	Mean + 2 S.D.	Range
2.13	0.17	2.47	1.8–2.5

Side to side difference: 0.05 ± 0.05 (range –0 to 0.2)

Peak to peak amplitude (μV)

Mean	S.D.	Mean – 2 S.D.	Range
27.6	11.2	5.2	10–50

Side to side proportion—smaller/larger = 0.88 ± 0.09 (range –0.7 to 1.0)

Spindler and Felsenthal Technique

Recording electrodes: A 3 cm bar electrode is used with the active electrode (A) 12 cm distal to the cathode on a line to the radial artery at the wrist. The reference electrode (R) is distal.

Ground electrode (G): Placement is between the stimulating and recording electrodes.

Stimulation point (S): The cathode (C) is placed just lateral to the biceps tendon at the distal elbow crease. The anode (A) is proximal.

Machine settings: Standard sensory settings are used.

Normal values (3) (30 subjects):

Onset latency (msec)

Mean	S.D.	Mean + 2 S.D.	Range
1.8	0.1	2.0	1.6–2.1

Side to side difference: 0.1 ± 0.1 msec (range 0.0–0.3)

Peak latency (msec)

Mean	S.D.	Mean + 2 S.D.	Range
2.3	0.1	2.5	2.2–2.6

Side to side difference: 0.1 ± 0.1 msec (range 0.0–0.3)

Peak to peak amplitude (μV)

Mean	S.D.	Mean − 2 S.D.	Range
24	7.2	9.6	12–50

Side to side proportion—smaller/larger = 0.89 ± 0.08 (range 0.7–1.0)

Helpful Hints

- When compared with the medial antebrachial cutaneous nerve study, the lateral antebrachial nerve generally has a larger amplitude.
- The cathode should be placed immediately next to the biceps tendon to obtain an optimal recording. Slight pressure may be necessary.

Notes

REFERENCES

1. Izzo KL, Aravabhumi S, Jafri A, et al: Medial and lateral antebrachial cutaneous nerves: standardization of technique, reliability and age effect on healthy subjects. *Arch Phys Med Rehabil* 1985; 66:592–597.
2. Kimura I, Ayyar DR: Sensory nerve conduction study in the medial antebrachial cutaneous nerve. *Tohoku J Exp Med* 1984; 142:461–466.
3. Spindler HA, Felsenthal G: Sensory conduction in the musculocutaneous nerve. *Arch Phys Med Rehabil* 1978; 59:20–23.

MEDIAL ANTEBRACHIAL CUTANEOUS SENSORY NERVE STUDY

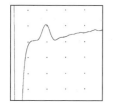

Typical waveform appearance

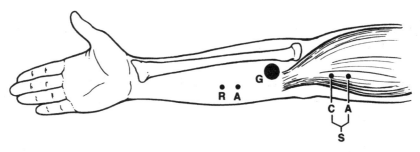

Izzo et al. Technique

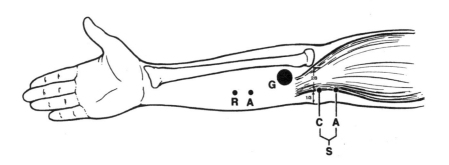

Kimura and Ayyar Technique

Electrode Placement

Izzo et al. Technique

Recording electrodes: A 3 cm bar electrode is placed on the anteromedial forearm with the active electrode (A) 14 cm distal to the stimulation point. The reference electrode (R) is distal.

Ground electrode (G): Placement is between the stimulating and recording electrodes.

Stimulation point (S): The cathode (C) is placed 5 cm proximal to the medial epicondyle. The anode (A) is proximal.

Machine settings: Standard sensory settings are used, with a sensitivity adequate to record 1 µV potentials. Averaging of eight responses was performed to derive the data presented.

Nerve fibers tested: C8 and T1 nerve roots, through the lower trunk, anterior division, and medial cord of the brachial plexus.

Normal values (1) (157 subjects—absent in 2) (skin temperature over the forearm 30–33 degrees Celsius):

Peak latency (msec)

Mean	S.D.	Mean + 2 S.D.	Range
2.7	0.2	3.1	2.1–3.3

Peak to peak amplitude (µV)

Mean	S.D.	Range	Lower Limit of Normal (5th Percentile)
11.4	5.2	3.0–32.0	5

Kimura and Ayyar Technique

Recording electrodes: The active electrode (A) is placed 10 cm distal to the cathode along a line connecting the cathode to the pisiform bone at the wrist. The reference electrode (R) is placed approximately 3 cm distally.

Ground electrode (G): Placement is between the stimulating and recording electrodes.

Stimulation point (S): The cathode (C) is placed medial to the biceps tendon at the junction of the medial third and lateral two thirds of a line connecting the biceps tendon to the medial epicondyle. The anode (A) is proximal.

Machine settings: Low frequency filter—32 Hz, High frequency filter—3.2 kHz

Normal values (2) (40 subjects):

Onset latency (msec)

Mean	S.D.	Mean + 2 S.D.	Range
1.54	0.17	1.88	1.2–2.0

Side to side difference: 0.05 ± 0.05 (range –0 to 0.2)

Peak latency (msec)

Mean	S.D.	Mean + 2 S.D.	Range
2.11	0.16	2.43	1.9–2.4

Side to side difference: 0.05 ± 0.04 (range –0 to 0.2)

Peak to peak amplitude (µV)

Mean	S.D.	Mean – 2 S.D.	Range
18.8	7.1	4.6	6–40

Side to side proportion—smaller/larger = 0.89 ± 0.10 (range –0.7 to 1.0)

Helpful Hints

- When compared with the medial antebrachial cutaneous nerve study, the lateral antebrachial nerve generally has a larger amplitude.
- Occasionally motor artifact can obscure the recording. Lowering the stimulus intensity may be helpful.
- Occasionally stimulus artifact can interfere with the recording. Rotating the anode or moving the ground electrode to the back of the forearm may be helpful.

Notes

REFERENCES

1. Izzo KL, Aravabhumi S, Jafri A, et al: Medial and lateral antebrachial cutaneous nerves: standardization of technique, reliability and age effect on healthy subjects. *Arch Phys Med Rehabil* 1985; 66:592–597.
2. Kimura I, Ayyar DR: Sensory nerve conduction study in the medial antebrachial cutaneous nerve. *Tohoku J Exp Med* 1984; 142:461–466.

ADDITIONAL READINGS/ALTERNATE TECHNIQUES

1. Pribyl R, You SB, Jantra P: Sensory nerve conduction velocity of medial antebrachial cutaneous nerve. *Electromyogr clin Neurophysiol* 1979; 19:41–46.
2. Reddy MP: Conduction studies of medial cutaneous nerve of forearm. *Arch Phys Med Rehabil* 1983; 64:209–211.
3. Ma DM, Liveson JA: *Nerve conduction handbook.* Philadelphia: FA Davis, 1983.

MEDIAN NERVE

MEDIAN SENSORY NERVE TO THE 2ND AND 3RD DIGITS

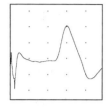

Typical waveform appearance

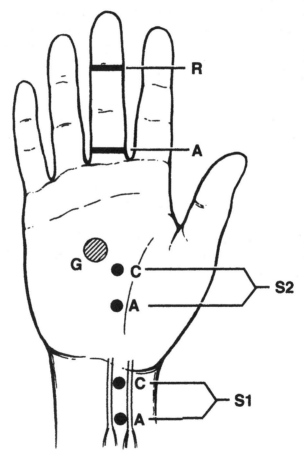

Electrode Placement

Active electrode (A): A ring or clip electrode is placed in contact with the radial and ulnar sides of the digit being tested, slightly distal to the base of the digit.

Reference electrode (R): A ring or clip electrode is placed in contact with the radial and ulnar sides of the digit being tested, 4 cm distal to the active electrode.

Ground electrode (G): Placement is on the dorsum of the hand.

Stimulation point 1 (S1): The subject is asked to straighten the fingers. The cathode (C) is placed 14 cm proximal to the active electrode over the median nerve at the wrist, between the tendons of the flexor carpi radialis and the palmaris longus (if the palmaris longus is absent, stimulation is applied slightly medial to the flexor carpi radialis tendon). The anode (A) is proximal.

Stimulation point 2 (S2): The cathode (C) is placed at the midpoint of the line from the active electrode to stimulation point 1. The anode (A) is proximal.

Machine settings: Sensitivity—20 µV/division, Low frequency filter—20 Hz, High frequency filter—2 kHz, Sweep speed—1 msec/division.

Nerve fibers tested: C6 (2nd digit) and C7 (3rd digit) nerve roots through the upper and middle trunks, anterior divisions, and lateral cord of the brachial plexus.

Normal values: (1) (258 subjects) (skin temperature over the dorsum of the hand 32 degrees Celsius or greater). The data presented are for the 3rd digit; the results for the 2nd digit are virtually identical.

Onset latencies (msec)

	Mean	S.D.	Mean + 2 S.D.	97th Percentile
S1	2.7	0.3	3.3	3.2
S2	1.4	0.2	1.8	1.8

Peak latencies (msec) (not all data were normally distributed—the mean + 2 S.D. values for S1 stimulation were derived through a logarithmic transformation)

	Mean	S.D.	Mean + 2 S.D.	97th Percentile
S1	3.4	0.3	4.1	4.0
S2	2.0	0.4	2.8	2.4

Onset to peak amplitude (μV) (the data were not normally distributed—the mean − 2 S.D. values were derived through a square root transformation). The data are divided into groups according to age and body mass index (BMI), kg/m^2 (see Appendix).

	Mean	S.D.	Mean − 2 S.D.	3rd Percentile
S1				
Age 19–49				
BMI < 24	51	19	19	
BMI ≥ 24	45	19	15	
Age 50–79				
BMI < 24	30	10	13	
BMI ≥ 24	24	10	8	
All subjects	41	20	10	14

Peak to peak amplitude (µV) (the data were not normally distributed—the mean – 2 S.D. values were derived through a square root transformation). The data are divided into groups according to age and body mass index (BMI), kg/m² (see Appendix).

	Mean	S.D.	Mean – 2 S.D.	3rd Percentile
S1				
Age 19–49				
BMI < 24	82	33	27	
BMI ≥ 24	69	31	19	
Age 50–79				
BMI < 24	47	16	18	
BMI ≥ 24	34	16	8	
All subjects	63	33	12	17

Area under the peak from onset to positive peak (nVsec) (the data were not normally distributed—the mean – 2 S.D. values were derived through a logarithmic transformation). The data are divided into groups according to age and body mass index (BMI), kg/m² (see Appendix).

	Mean	S.D.	Mean – 2 S.D.	3rd Percentile
S1				
Age 19–49				
BMI < 24	65	31	22	
BMI ≥ 24	53	27	18	
Age 50–79				
BMI < 24	37	15	15	
BMI ≥ 24	27	13	9	
All subjects	49	29	13	14

Rise time (msec) (not all data were normally distributed—the mean + 2 S.D. value for S1 was derived through a logarithmic transformation)

		Mean	S.D.	Mean + 2 S.D.	97th Percentile
S1	0.7	0.1		1.0	1.0
S2	0.6	0.3		1.2	0.8

Duration from onset to positive peak (msec)

	Mean	S.D.	Mean + 2 S.D.	97th Percentile
S1	2.1	0.4	2.9	2.9
S2	1.8	0.5	2.8	2.7

Acceptable Differences

The upper limit of normal increase in onset and peak latency from one side to the other is 0.4 msec.

The upper limit of normal decrease in onset to peak amplitude from one side to the other is 50%.

The upper limit of normal decrease in peak to peak amplitude from one side to the other is 55%.

The upper limit of normal decrease in area from one side to the other is 63%.

The upper limit of normal percentage of the S1 onset latency attributable to the wrist-to-palm segment (S1 minus S2/S1) is 58%.

The upper limit of normal percentage of the S1 peak latency attributable to the wrist-to-palm segment (S1 minus S2/S1) is 50%.

The upper limit of normal increase in onset to peak and peak to peak amplitude from S1 to S2 is approximately 50%

The upper limit of normal increase in latency from digit 2 to digit 3 is 0.4 msec for onset latency and 0.3 msec for peak latency.

The upper limit of normal increase in latency from digit 3 to digit 2 is 0.2 msec for onset and peak latency.

The upper limit of normal decrease in onset to peak amplitude from one digit to the other is approximately 44–48%.

The upper limit of normal decrease in peak to peak amplitude from one digit to the other is approximately 50%.

The upper limit of normal decrease in area from one digit to the other is approximately 40–50%.

The upper limit of normal difference in median and ulnar 14 cm antidromic latency of the same hand is generally thought to be 0.4–0.5 msec (2–4).

Helpful Hints

- After applying the clip electrodes to the digits, they should be rotated from side to side to help spread the electrode paste.

- Volume conduction from the muscles of the hand may be seen as a motor wave, usually slightly after the sensory response. If this is obscuring the sensory recording, the active and recording electrodes may need to be repositioned slightly more distal on the digit. In the case of an absent sensory response, the examiner may misidentify the motor response as a delayed sensory recording.

- If there is doubt about whether the observed recording is truly a sensory response, the recording and stimulating electrodes may need to be reversed to perform an orthodromic recording. If this is done, the evoked response amplitudes can be expected to be smaller than with digital recording. Such responses may need to be averaged or even recorded with near nerve needle recording.

- If the skin of the palm is thick, it can make the S2 response difficult to elicit. Mild abrasion of the skin or needle stimulation may be needed.

- To avoid contamination of the response by contact of the clips with the adjacent fingers, a small roll of gauze may be placed between the digits to hold them apart.

- In persons with short hands the normal 7 cm S2 site may be situated over the wrist rather than in the palm. In such cases S2 may be moved more distally. This will still allow amplitude comparison from S1 to S2, but not latency comparison. Alternatively, S1 may be placed 12 cm proximal to the active electrode and S2 would be at 6 cm. This will still allow a latency ratio comparison.

- Stimulation can be applied across the wrist in 1 cm increments in the so-called "inching" technique. Normally as the stimulus is applied 1 cm more distally the latency increases by 0.16–0.21 msec per 1 cm increment. Greater incremental latency changes can help to localize focal slowing (5). Upper limits of normal increase in incremental latencies of 0.4 msec and 0.5 msec, or a doubling of the adjacent segments' latencies, have been recommended (6–8). Ross and Kmura recommend that inching be done not to diagnose carpal tunnel syndrome, but to confirm the site of slowing (8).

Notes

REFERENCES

1. Buschbacher RM: Median 14 cm and 7 cm antidromic sensory studies to digits 2 and 3. *Am J Phys Med Rehabil* 1999; 78:S53–S62.
2. Felsenthal G: Median and ulnar distal motor and sensory latencies in the same normal subject. *Arch Phys Med Rehabil* 1977; 58:297–302.
3. Felsenthal G, Spindler H: Palmar conduction time of median and ulnar nerves of normal subjects and patients with carpal tunnel syndrome. *Am J Phys Med Rehabil* 1979; 58:131–138.
4. Monga TN, Shanks GL, Poole BJ: Sensory palmar stimulation in the diagnosis of carpal tunnel syndrome. *Arch Phys Med Rehabil* 1985; 66:598–600.
5. Kimura J: The carpal tunnel syndrome: localization of conduction abnormalities within the distal segment of the median nerve. *Brain* 1979; 102:619–635.

6. Nathan PA, Keniston RC, Meadows KD, Lockwood RS: Predictive value of nerve conduction measurements at the carpal tunnel. *Muscle Nerve* 1993; 16:1377–1382.
7. Nathan PA, Meadows KD, Doyle LS: Sensory segmental latency values of the median nerve for a population of normal individuals. *Arch Phys Med Rehabil* 1988; 69:499–501.
8. Ross MA, Kimura J: AAEM case report #2: the carpal tunnel syndrome. *Muscle Nerve* 1995; 18:567–573.

ADDITIONAL READINGS/ALTERNATE TECHNIQUES

1. Melvin JL, Schuchmann JA, Lanese RR: Diagnostic specificity of motor and sensory nerve conduction variables in the carpal tunnel syndrome. *Arch Phys Med Rehabil* 1973; 54:69–74.
2. Wongsam PE, Johnson EW, Weinerman JD: Carpal tunnel syndrome: use of palmar stimulation of sensory fibers. *Arch Phys Med Rehabil* 1983; 64:16–19.
3. Hennessey WJ, Falco FJE, Braddom RL: Median and ulnar nerve conduction studies: normative data for young adults. *Arch Phys Med Rehabil* 1994; 75:259–264.
4. Hennessey WJ, Falco FJE, Goldberg G, Braddom RL: Gender and arm length: influence on nerve conduction parameters in the upper limb. *Arch Phys Med Rehabil* 1994; 75:265–269.
5. Falco FJE, Hennessey WJ, Braddom RL, Goldberg G: Standardized nerve conduction studies in the upper limb of the healthy elderly. *Am J Phys Med Rehabil* 1992; 71:263–271.
6. Salerno DF, Franzblau A, Werner RA, et al: Median and ulnar nerve conduction studies among workers: normative values. *Muscle Nerve* 1998; 21:999–1005.

MEDIAN PALMAR CUTANEOUS SENSORY NERVE STUDY

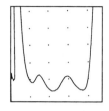

Typical waveform appearance

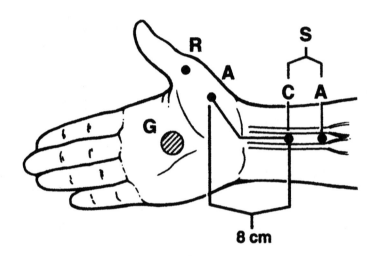

Electrode Placement

Set-up is as for median motor nerve conduction to the abductor pollicis brevis.

Active electrode (A): Placement is halfway between the midpoint of the distal wrist crease and the 1st metacarpophalangeal joint.

Reference electrode (R): Placement is slightly distal to the 1st metacarpophalangeal joint.

Ground electrode (G): Placement is on the dorsum of the hand. If stimulus artifact interferes with the recording, the ground may be placed near the active electrode, between this electrode and the cathode.

Stimulation point (S): The cathode (C) is placed 8 cm proximal to the active electrode, in a line measured first to the midpoint of the distal wrist crease and then to a point slightly ulnar to the tendon of the flexor carpi radialis. The anode (A) is proximal.

Machine settings: Sensitivity—10 µV/division, Low frequency filter—20 Hz, High frequency filter 2 kHz, Sweep speed—1 msec/division

Nerve fibers tested: C6 nerve root through the upper trunk, anterior division, and lateral cord of the brachial plexus.

This branch leaves the median nerve above the carpal tunnel.

Normal values (10 subjects):

Peak latency (msec)

Mean	S.D.	Mean + 2 S.D.	Range
1.54	0.08	1.70	1.38–1.70

Peak to peak amplitude (µV)

Mean	S.D.
9.03	10.58

Helpful Hints

- There are two small amplitude negative waves preceding the compound motor action potential of the abductor pollicis brevis. Both waves should be visualized using the technique described above. This will enable the examiner to correctly identify the sensory nerve action potential of the median palmar cutaneous nerve, which is the first of the two small amplitude negative waves. The second negative wave is a far field recording of either the junctional potential of the median digital nerve entering the thumb or that of a fixed neural generator in the palm.

- The cathode may be repositioned laterally to attempt to selectively depolarize the median palmar cutaneous nerve. Using a short duration stimulus may decrease the recording interference from the stimulus artifact. Using a submaximal stimulus intensity from that used to record the motor response may also be helpful. Repositioning of the active electrode proximally may also be helpful. The amplitude will be larger with more proximal placement and therefore may be more easily detected in some individuals. Care should be taken not to place the active electrode too far medially, as the sensory recording might be obscured by depolarization of the main median trunk.

Notes

REFERENCE

Bergeron JW, Braddom RL: Palmar cutaneous nerve recording and clarification of median premotor potential generators. *Am J Phys Med Rehabil* 1998; 77:399–406.

POSTERIOR ANTEBRACHIAL CUTANEOUS SENSORY NERVE STUDY

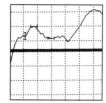

Typical waveform appearance

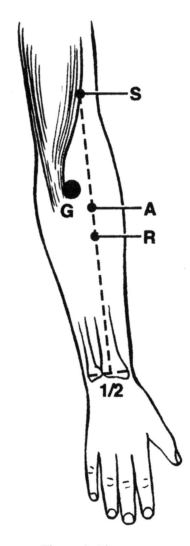

Electrode Placement

Recording electrodes: Placement is along a line from the stimulation point to the mid-dorsum of the wrist (midway between the radial and ulnar styloid processes). The active electrode (A) is placed approximately 12 cm distal to the stimulating electrode. The reference electrode (R) is 3 cm distal.

Ground electrode (G): Placement is between the stimulating and recording electrodes.

Stimulation point (S): Placement is just above the lateral epicondyle, between the biceps and triceps, slightly closer to the latter muscle, at the border of the lateral head of the triceps.

Nerve fibers tested: C5–C8 nerve roots, through the upper, middle, and lower trunks, posterior divisions, and posterior cord of the brachial plexus, and then through the radial nerve.

Normal values (1) (22 subjects) (distance 10.0–14.0 cm, room temperature 23–26 degrees Celsius):

Onset latency (msec)

Mean	S.D.	Mean + 2 S.D.	Range
1.9	0.3	2.5	1.5–2.4

Peak to peak amplitude (µV)

Mean	S.D.	Range
8.6	3.9	5.0–20.0

Helpful Hints

- This nerve supplies sensation to the skin of the lateral arm and elbow and dorsal forearm to the wrist.
- This test is best performed with the forearm pronated.

- The waveform may be difficult to record because of motor artifact. Lower intensity stimulation may be necessary. Adjustment of the stimulating and recording electrodes may be needed as well.
- Initially stimulation should be performed 1/2–2 cm directly above the lateral epicondyle. If no response is obtained, the stimulator should be moved anteriorly or posteriorly.

Notes

REFERENCE

1. Ma DM, Liveson JA: *Nerve conduction handbook.* Philadelphia: FA Davis, 1983.

RADIAL SENSORY NERVE STUDY TO THE BASE OF THE THUMB

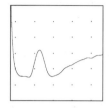

Typical waveform appearance

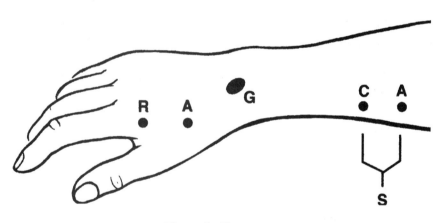

Electrode Placement

Active electrode (A): Placement is over the superficial radial nerve where it can be felt to cross over the tendon of the extensor pollicis longus at the wrist.

Reference electrode (R): Placement is slightly proximal to the 2nd metacarpal head on the lateral side of this bone.

Ground electrode (G): Placement is between the stimulating and recording electrodes.

Stimulation point (S): The cathode (C) is placed on the radial side of the forearm 10 cm proximal to the active electrode (wrist in neutral, thumb slightly adducted). The anode (A) is proximal.

Machine settings: Sensitivity—10 µV/division, Low frequency filter—2 Hz, High frequency filter—2 kHz, Sweep speed—2 msec/division.

Nerve fibers tested: C6 nerve root, through the upper trunk, posterior division, and posterior cord of the brachial plexus.

Normal values (1) (49 subjects) (skin temperature over the nerve 31 degrees Celsius or greater)

Onset latency (msec)

Mean	S.D.	Mean + 2 S.D.
1.8	0.3	2.4

Peak latency (msec)

Mean	S.D.	Mean + 2 S.D.
2.3	0.4	3.1

Peak to peak amplitude (µV)

Mean	S.D.	Range
31	20	13–60

Helpful Hint

- Asking the subject to actively extend the thumb may help in palpating and localizing the nerve.

Notes

REFERENCE

1. Mackenzie K, DeLisa JA: Distal sensory latency measurement of the superficial radial nerve in normal adult subjects. *Arch Phys Med Rehabil* 1981; 62:31–34.

ADDITIONAL READINGS/ALTERNATE TECHNIQUES

1. Downie AW, Scott TR: An improved technique for radial nerve conduction studies. *J Neurol Neurosurg Psychiatry* 1967; 30:332–336.
2. Ma DM, Liveson JA: *Nerve conduction handbook.* Philadelphia: FA Davis, 1983.

3. Chang CW, Oh SJ: Sensory nerve conduction study in forearm segment of superficial radial nerve: standardization of technique. *Electromyogr clin Neurophysiol* 1990; 30:349–351.
4. Hoffman MD, Mitz M, Luisi M, Melville BR: Paired study of the dorsal cutaneous ulnar and superficial radial sensory nerves. *Arch Phys Med Rehabil* 1988; 69:591–594.

ULNAR NERVE

ULNAR DORSAL CUTANEOUS SENSORY NERVE STUDY

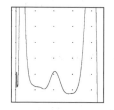

Typical waveform appearance

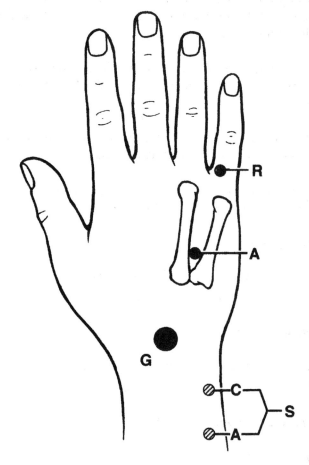

Electrode Placement

Hoffman et al. Technique

Recording electrodes: A 3 cm bar is placed over the ulnar portion of the dorsal hand with the active electrode (A) proximal and the reference electrode (R) distal.

Ground electrode (G): Placement is between the stimulating and recording electrodes.

Stimulation point (S): The cathode (C) is placed 14 cm proximal to the active electrode between the ulna and the tendon of the flexor carpi ulnaris. The anode (A) is proximal.

Machine settings: Sensitivity—20–50 µV/division, Low frequency filter—32 Hz, High frequency filter—3.2 kHz, Sweep speed—1 msec/division

Nerve fibers tested: C8 nerve root, through the lower trunk, anterior division, and medial cord of the brachial plexus.

Normal values: (1) (33 subjects) (skin temperature on the dorsum of the hand between 31 and 36 degrees Celsius)

Onset latency (msec)

Mean	S.D.	Mean + 2 S.D.
2.2	0.3	2.8

Peak latency (msec)

Mean	S.D.	Mean + 2 S.D.
2.8	0.5	3.8

Baseline to peak amplitude (µV)

Mean	S.D.
24	17

Jabre Technique

Active electrode (A): Placement is over the proximal point of the "V" formed by the 4th and 5th metacarpal bones on the dorsum of the hand. The nerve can often be palpated as it crosses the 4th and 5th metacarpals. If so, this site is chosen.

Reference electrode (R): Placement is at the base of the 5th digit.

Ground electrode (G): Placement is between the stimulating and recording electrodes.

Stimulation point (S): The cathode (C) is placed 8 cm proximal to the active electrode over the bony ulna or in the space between the ulna and the flexor carpi ulnaris (whichever position gives a better baseline). The anode (A) is proximal. The patient lies supine, with the arm alongside the body and the hand pronated. When muscle relaxation is difficult to obtain in this position, the hand is supinated.

Machine settings: Routine sensory settings are used.

Normal values (2) (30 subjects) (no temperature control):

Latency (msec)

Mean	S.D.	Mean + 2 S.D.
2.0	0.3	2.6

Amplitude (μV)

Mean	S.D.
29	6

Kim et al. Technique

Recording electrodes: A 3 cm bar electrode is used with the reference electrode (R) over the 5th metacarpophalangeal joint and the active electrode (A) proximal over the 5th metacarpal.

Ground electrode (G): Placement is between the stimulating and recording electrodes.

Stimulation point (S): The cathode (C) is placed 8–10 cm proximal to the ulnar styloid between the ulna and the tendon of the flexor carpi ulnaris. The anode (A) is proximal. The subjects are seated, arm supported, forearm pronated, wrist in neutral, and fingers slightly flexed. The distance from stimulating to recording electrodes is approximately 11–12 cm.

Machine settings: Standard sensory settings are used.

Normal values (3) (33 subjects) (no temperature control):

Peak latency

Mean	S.D.	Mean + 2 S.D.
2.1	0.3	2.7

Peak to peak amplitude

Mean	S.D.
24.2	10.8

Only 2 nerves had amplitudes of less than 10 µV

Helpful Hints

- Anomalous innervation to this area may be present (2,3).
- Jabre's study used somewhat unusual subject inclusion criteria. In 11% of his subjects there was a greater than 50% side to side difference in amplitude, and he omitted the data for these persons, assuming that they had asymptomatic nerve pathology.
- This sensory recording is often obscured by motor artifact. Submaximal stimulation can be useful if this is the case.
- Kim and coworkers reported that if difficulty was encountered in obtaining a response, the following maneuvers can be used: (1) stimulator can be moved 2–3 cm more proximal or distal, (2) an orthodromic technique can be used, and (3) the radial and musculocutaneous nerves can be stimulated to test for an anomalous innervation pattern.
- Hoffman and coworkers reported that supinating the forearm produced better recordings.

Notes

REFERENCES

1. Hoffman MD, Mitz M, Luisi M, Melville BR: Paired study of the dorsal cutaneous ulnar and superficial radial sensory nerves. *Arch Phys Med Rehabil* 1988; 69:591–594.
2. Jabre JF: Ulnar nerve lesions at the wrist: new technique for recording from the sensory dorsal branch of the ulnar nerve. *Neurology* 1980; 30:873–876.
3. Kim DJ, Kalantri A, Guha S, Wainapel SF: Dorsal cutaneous ulnar nerve conduction. *Arch Neurol* 1981; 38:321–322.

ADDITIONAL READING/ALTERNATE TECHNIQUE

1. Young SH, Kalantri A: Dorsal ulnar cutaneous nerve conduction studies in an asymptomatic population (abstract). *Arch Phys Med Rehabil* 1998; 79:1166.

ULNAR SENSORY NERVE STUDY TO THE 5TH DIGIT

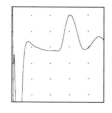

Typical waveform appearance

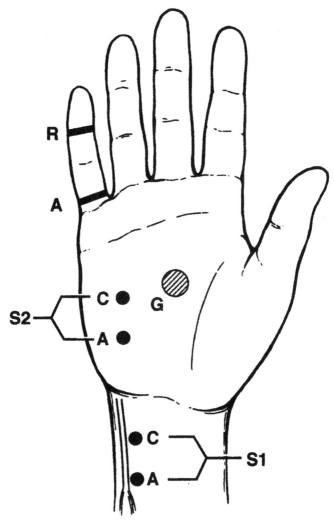

Electrode Placement

Active electrode (A): A ring or clip electrode is placed in contact with the radial and ulnar sides of the 5th digit, slightly distal to the base of the digit.

Reference electrode (R): A ring or clip electrode is placed in contact with the radial and ulnar sides of the 5th digit, 4 cm distal to the active electrode (or in small fingers as far distally as possible).

Ground electrode (G): Placement is on the dorsum of the hand.

Stimulation point 1 (S1): The subject is asked to straighten the fingers. The cathode (C) is placed 14 cm proximal to the active electrode over the ulnar nerve at the wrist, slightly radial to the tendon of the flexor carpi ulnaris. The anode (A) is proximal.

Stimulation point 2 (S2): The cathode (C) is placed at the midpoint of the line from the active electrode to stimulation point 1. The anode (A) is proximal.

Machine settings: Sensitivity—20 µV/division, Low frequency filter—20 Hz, High frequency filter—2 kHz, Sweep speed—1 msec/division.

Nerve fibers tested: C8 nerve root, through the lower trunk, anterior division, and medial cord of the brachial plexus.

Normal values (1) (258 subjects) (skin temperature over the dorsum of the hand 32 degrees Celsius or greater):

Onset latencies (msec)

	Mean	S.D.	Mean + 2 S.D.	97th Percentile
S1	2.6	0.2	3.0	3.1
S2	1.4	0.2	1.8	1.8

Peak latencies (msec) (the data were not normally distributed—the mean + 2 S.D. values were derived through a logarithmic transformation).

	Mean	S.D.	Mean + 2 S.D.	97th Percentile
S1	3.4	0.3	4.1	4.0
S2	2.0	0.2	2.5	2.5

Onset to peak amplitude (µV) (the data were not normally distributed—the mean − 2 S.D. values were derived through a square root transformation). The data are divided into groups according to age and body mass index (BMI), kg/m² (see Appendix).

	Mean	S.D.	Mean − 2 S.D.	3rd Percentile
S1				
Age 19–49				
BMI < 24	44	17	14	
BMI ≥ 24	36	16	11	
Age 50–79				
BMI < 24	27	11	10	
BMI ≥ 24	20	9	5	
All subjects	33	17	6	10

Peak to peak amplitude (µV) (the data were not normally distributed—the mean − 2 S.D. values were derived through a square root transformation). The data are divided into groups according to age and body mass index (BMI), kg/m² (see Appendix).

	Mean	S.D.	Mean − 2 S.D.	3rd Percentile
S1				
Age 19–49				
BMI < 24	70	36	13	
BMI ≥ 24	54	31	8	
Age 50–79				
BMI < 24	37	13	13	
BMI ≥ 24	27	15	4	
All subjects	50	32	4	9

Area under the peak from onset to positive peak (nVsec) (the data were not normally distributed—the mean − 2 S.D. values were derived through a logarithmic transformation). The data are divided into groups according to age and body mass index (BMI), kg/m² (see Appendix).

	Mean	S.D.	Mean − 2 S.D.	3rd Percentile
S1				
Age 19–49				
BMI < 24	53	29	14	
BMI ≥ 24	38	22	10	
Age 50–79				
BMI < 24	28	11	12	
BMI ≥ 24	22	14	6	
All subjects	37	24	8	9

Rise time (msec) (the data were not normally distributed—the mean + 2 S.D. values were derived through a logarithmic transformation).

	Mean	S.D.	Mean + 2 S.D.	97th Percentile
S1	0.8	0.2	1.2	1.1
S2	0.6	0.1	0.9	0.9

Duration from onset to positive peak (msec)

	Mean	S.D.	Mean + 2 S.D.	97th Percentile
S1	1.9	0.4	2.7	2.7
S2	1.7	0.3	2.3	2.5

Acceptable Differences

The upper limit of normal increase in onset latency from one side to the other is 0.3 msec.

The upper limit of normal increase in peak latency from one side to the other is 0.4 msec.

The upper limit of normal decrease in onset to peak amplitude from one side to the other is 53%.

The upper limit of normal decrease in peak to peak amplitude from one side to the other is 64%.

The upper limit of normal decrease in area from one side to the other is 65%.

The upper limit of normal percentage of the S1 onset latency attributable to the wrist-to-palm segment (S1 minus S2/S1) is 54%.

The upper limit of normal percentage of the S1 peak latency attributable to the wrist-to-palm segment (S1 minus S2/S1) is 47%.

The upper limit of normal increase in onset to peak amplitude from S1 to S2 is 71%.

The upper limit of normal increase in peak to peak amplitude from S1 to S2 is 60%.

Helpful Hints

- After applying the clip electrodes to the digits, they should be rotated from side to side to help spread the electrode paste.

- Volume conduction from the muscles of the hand may be seen as a motor wave, usually slightly after the sensory response. If this is obscuring the sensory recording, the active and recording electrodes may need to be repositioned slightly more distal on the digit. In the case of an absent sensory response, the examiner may misidentify the motor response as a delayed sensory recording.

- If there is doubt about whether the observed recording is truly a sensory response, the recording and stimulating electrodes may need to be reversed to perform an orthodromic recording. If this is done, the evoked response amplitudes can be expected to be smaller than with digital recording. Such responses may need to be averaged or even recorded with near nerve needle recording.

- If the skin of the palm is thick, it may make the S2 response difficult to elicit. Mild abrasion of the skin or needle stimulation may be needed.

- To avoid contamination of the response by contact of the clips with the adjacent finger, a small roll of gauze may be placed between the digits to hold them apart.

- In persons with short hands the normal 7 cm S2 site may be situated over the wrist rather than in the palm. In such cases S2 may be moved more distally. This will still allow amplitude comparison from S1 to S2, but not latency comparison. Alternatively, S1 may be placed 12 cm proximal to the active electrode and S2 would be at 6 cm. This will still allow a latency ratio comparison.

- The sensory studies are usually performed only for relatively short distances. This is due to the fact that sensory compound action potentials are particularly sensitive to phase cancellation, causing a rapid decrease in amplitude with increasing distance from the recording site. One study has presented the data for sensory recording with more proximal stimulation for 20 subjects (averaging used). The mean + 2 S.D. value for latency change for a 10 cm segment across the elbow (see ulnar nerve motor study to the abductor digiti minimi) was 1.8 msec for onset latency and 1.9 msec for peak latency. Side to side difference (mean + 2 S.D.) in latency change across this segment was 0.6 msec for onset latency and 0.3 msec for peak latency. The upper limit of normal decrement of peak to peak amplitude was 74% (range 12–72%) across the wrist to below elbow segment and 41% (range 0–50%) for the below to above elbow segment. The lower limits of normal nerve conduction velocity (calculated for onset latency) were 59 m/sec below the elbow and 50 m/sec across the elbow) (2).

Notes _____

REFERENCES

1. Buschbacher RM: Ulnar 14 cm and 7 cm antidromic sensory studies to the 5th digit. *Am J Phys Med Rehabil* 1999; 78:S63–S68.
2. Felsenthal G, Freed MJ, Kalafut R, Hilton EB: Across-elbow ulnar nerve sensory conduction technique. *Arch Phys Med Rehabil* 1989; 70:668–672.

ADDITIONAL READINGS/ALTERNATE TECHNIQUES

1. Stowell ED, Gnatz SM: Ulnar palmar cutaneous nerve and hypothenar sensory conduction studies. *Arch Phys Med Rehabil* 1992; 73:842–846.
2. Falco FJE, Hennessey WJ, Braddom RL, Goldberg G: Standardized nerve conduction studies in the upper limb of the healthy elderly. *Am J Phys Med Rehabil* 1992; 71:263–271.
3. Hennessey WJ, Falco FJE, Braddom RL: Median and ulnar nerve conduction studies: normative data for young adults. *Arch Phys Med Rehabil* 1994; 75:259–264.
4. Hennessey WJ, Falco FJE, Goldberg G, Braddom RL: Gender and arm length: influence on nerve conduction parameters in the upper limb. *Arch Phys Med Rehabil* 1994; 75:265–269.
5. Salerno DF, Franzblau A, Werner RA, et al: Median and ulnar nerve conduction studies among workers: normative values. *Muscle Nerve* 1998; 21:999–1005.

COMPARATIVE STUDIES

MEDIAN AND RADIAL SENSORY NERVES TO 1ST DIGIT

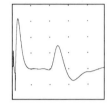

Typical waveform appearance

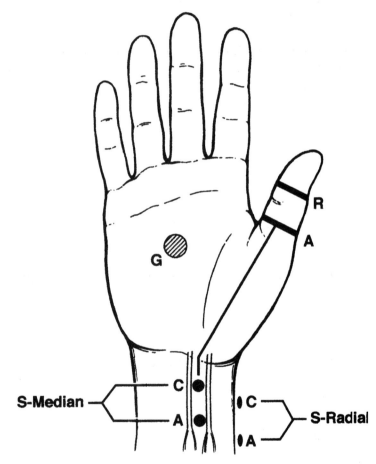

Electrode Placement

Active electrode (A): A ring electrode is placed on the 1st metacarpophalangeal joint. (1)

Reference electrode (R): A ring electrode is placed on the interphalangeal joint. (1)

Ground electrode (G): Placement is on the dorsum of the hand.

Stimulation point (s-radial): The cathode (C) is placed over the radial nerve as it is palpated on the lateral radius 10 cm proximal to the active electrode. The anode (A) is proximal.

Stimulation point (s-median): The cathode (C) is placed 10 cm proximal to the active electrode, in a line measured first to the midpoint of the distal wrist crease and then to a point slightly ulnar to the tendon of the flexor carpi radialis. The anode (A) is proximal.

Machine settings: Low frequency filter—20 Hz, High frequency filter—10 kHz (1)

Nerve fibers tested: Median: C6 nerve root through the upper trunk, anterior division, and lateral cord of the brachial plexus. Radial: C6 nerve root through the upper trunk, posterior division, and posterior cord of the brachial plexus.

Normal values (2) (78 subjects) (skin temperature over the proximal palm 33.5–34.5 degrees Celsius):

Peak latency (msec)

	Mean	S.D.	Mean + 2 S.D.	Range
Median	2.5	0.2	2.9	2.0–2.9
Radial	2.4	0.2	2.8	1.9–2.8

Baseline to peak amplitude (μV)

	Mean	S.D.
Median	30	2
Radial	12	1

Acceptable Difference

The upper limit of normal increase in median versus radial latency between the two nerve studies has been reported to be 0.3–0.4 msec (1–3).

Notes

REFERENCES

1. Pease WS, Cannell CD, Johnson EW: Median to radial latency difference test in mild carpal tunnel syndrome. *Muscle Nerve* 1989; 12:905–909.
2. Johnson EW, Sipski M, Lammertse T: Median and radial sensory latencies to digit I: normal values and usefulness in carpal tunnel syndrome. *Arch Phys Med Rehabil* 1987; 68; 140–141.
3. Cho DS, MacLean IC: Comparison of normal values of median, radial, and ulnar sensory latencies. *Muscle Nerve* 1984; 7:575.

ADDITIONAL READINGS/ALTERNATE TECHNIQUES

1. DiBenedetto M, Mitz M, Klingbeil GE, Davidoff D: New criteria for sensory nerve conduction especially useful in diagnosing carpal tunnel syndrome. *Arch Phys Med Rehabil* 1986; 67:586–589.
2. Carroll GJ: Comparison of median and radial nerve sensory latencies in the electrophysiological diagnosis of carpal tunnel syndrome. *Electroencephalogr clin Neurophysiol* 1987; 68:101–106.
3. Jackson DA: Clifford JC: Electrodiagnosis of mild carpal tunnel syndrome. *Arch Phys Med Rehabil* 1989; 70:199–204.
4. Falco FJE, Hennessey WJ, Braddom RL, Goldberg G: Standardized nerve conduction studies in the upper limb of the healthy elderly. *Am J Phys Med Rehabil* 1992; 71:263–271.
5. Hennessey WJ, Falco FJE, Goldberg G, Braddom RL: Gender and arm length: influence on nerve conduction parameters in the upper limb. *Arch Phys Med Rehabil* 1994; 75:265–269.
6. Kothari MJ, Rutkove SB, Caress JB, et al: Comparison of digital sensory studies in patients with carpal tunnel syndrome. *Muscle Nerve* 1995; 18:1272–1276.

MEDIAN AND ULNAR MIXED NERVE STUDIES

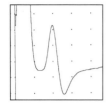

Typical waveform appearance

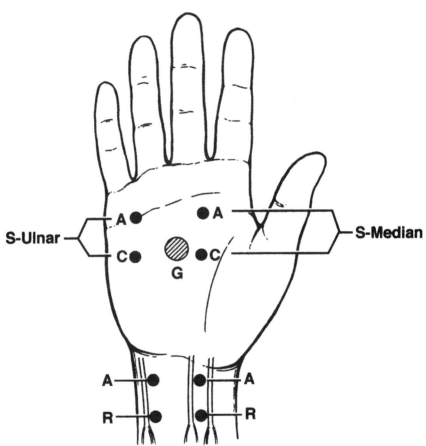

Electrode Placement

Recording electrodes (median): A 3 cm bar electrode is placed with the active electrode (A) at the proximal wrist crease between the tendons of the flexor carpi radialis and palmaris longus. If the palmaris longus is absent, the electrode is placed slightly ulnar to the tendon of the flexor carpi radialis. The reference electrode (R) is proximal.

Recording electrodes (ulnar): A 3 cm bar electrode is placed with the active electrode (A) at the proximal wrist crease slightly radial to the tendon of the flexor carpi ulnaris. The reference electrode (R) is proximal.

Ground electrode (G): Placement is on the dorsum of the hand.

Stimulation point (s-median): The cathode (C) is placed 8 cm distal to the active electrode in the mid-palm. The anode (A) is placed distally.

Stimulation point (s-ulnar): The cathode (C) is placed 8 cm distal to the active electrode in the lateral palm between the tendons of the flexors of the 4th and 5th digits.

Machine settings: Sensitivity—20 µV/division, Low frequency filter—20 Hz, High frequency filter—2 kHz, Sweep speed—1 msec/division.

Normal values (1) (248 subjects): (skin temperature over the dorsum of the hand at least 32 degrees Celsius):

Onset latencies (msec)

	Mean	*S.D.*	*Mean + 2 S.D.*	*97th Percentile*
Median	1.6	0.2	2.0	2.0
Ulnar	1.6	0.2	2.0	1.9

Peak latencies (msec)

	Mean	*S.D.*	*Mean + 2 S.D.*	*97th Percentile*
Median	2.1	0.2	2.5	2.4
Ulnar	2.1	0.2	2.5	2.4

Onset to peak amplitude (µV), peak to peak amplitude (µV), and area (nVsec-onset to positive peak). The means and standard deviations were derived independently for subgroups divided by age, sex, and body mass index (BMI—kg/m^2) (see Appendix). They are listed at the end of this section for completeness. The lower limits of normal differed only for women under age 50 for the median nerve and for women under age 30 with a BMI less than 24 kg/m^2 for the ulnar nerve. Therefore, the lower limits of normal are divided into two groups for each nerve study. The results were not distributed in a Gaussian fashion, thus the reader is cautioned against performing the mean − 2 S.D. calculation. The lower limits of normal are presented as the 3rd percentile of observed values.

Lower limits of normal (3rd percentile)

	Onset-to-Peak Amplitude	Peak-to-Peak Amplitude	Area
Median			
Women under 50	27	32	17
All others	15	14	10
Ulnar			
Women under 30 with BMI under 24	22	33	14
All others	6	6	4

Rise time (msec)

	Mean	S.D.	Mean + 2 S.D.	97th Percentile
Median	0.5	0.1	0.7	0.6
Ulnar	0.5	0.1	0.7	0.7

Duration from onset to positive peak (msec)

	Mean	S.D.	Mean + 2 S.D.	97th Percentile
Median	1.2	0.3	1.8	1.5
Ulnar	1.2	0.2	1.6	1.5

Acceptable Differences

The upper limit of normal increase in onset latency from one side to the other for the median and ulnar nerves is 0.3 msec.

The upper limit of normal increase in peak latency from one side to the other is 0.3 msec for the median nerve and 0.4 msec for the ulnar nerve.

The upper limit of normal decrease in onset to peak amplitude from one side to the other is 64% for the median nerve and 73% for the ulnar nerve.

The upper limit of normal decrease in peak to peak amplitude from one side to the other is 64% for the median nerve and 72% for the ulnar nerve.

The upper limit of normal decrease in area from one side to the other is approximately 59% for the median nerve and 73% for the ulnar nerve.

The upper limit of normal increase in onset or peak latency of one nerve versus the other is 0.3 msec.

Helpful Hints

- If stimulus is applied fairly distally in the palm, a more purely sensory response is recorded.
- If the subject's hand is small and the cathode is placed too distally to perform the technique, the bar electrode may need to be repositioned more proximally.

Notes

REFERENCE

1. Buschbacher RM: Mixed nerve conduction studies of the median and ulnar nerves. *Am J Phys Med Rehabil* 1999; 78:S69–S74.

ADDITIONAL READINGS/ALTERNATE TECHNIQUES

1. Eklund G: A new electrodiagnostic procedure for measuring sensory nerve conduction across the carpal tunnel. *Upsala J Med Sci* 1975; 80:63–64.
2. Daube JR: Percutaneous palmar median nerve stimulation for carpal tunnel syndrome. *Electroencephalogr clin Neurophysiol* 1977; 43:139–140.
3. Stowell ED, Gnatz SM: Ulnar palmar cutaneous nerve and hypothenar sensory conduction studies. *Arch Phys Med Rehabil* 1992; 73:842–846.
4. Daube JR, Stevens JC: The electrodiagnosis of carpal tunnel syndrome (a reply). *Muscle Nerve* 1993; 16:798.
5. Stevens JC: AAEM minimonograph #26: the electrodiagnosis of carpal tunnel syndrome. *Muscle Nerve* 1997; 20:1477–1486.

Mean and Standard Deviation Data for Onset-to-Peak Amplitude (μV), Peak-to-Peak Amplitude (μV), and Area (nVsec).

	Onset-to-Peak Amplitude Mean (SD)	Peak-to-Peak Amplitude Mean (SD)	Area Mean (SD)
Median			
Females			
BMI < 28			
Age 19–49	112 (49)	113 (49)	58 (26)
Age 50–79	48 (34)	52 (32)	28 (18)

BMI ≥ 28			
Age 19–49	92 (38)	98 (38)	46 (18)
Age 50–79	47 (22)	50 (25)	25 (12)
Males			
BMI < 28			
Age 19–49	68 (42)	78 (48)	41 (25)
Age 50–79	42 (18)	48 (26)	25 (12)
BMI ≥ 28			
Age 19–49	52 (30)	58 (37)	28 (15)
Age 50–79	41 (23)	41 (21)	24 (11)
All subjects	75 (47)	80 (48)	41 (24)

Ulnar

Females			
BMI < 24			
Age 19–29	53 (18)	56 (24)	32 (12)
Age 30–59	32 (17)	38 (19)	18 (9)
Age 60–79	19 (7)	24 (11)	11 (5)
BMI ≥ 24			
Age 19–29	32 (16)	37 (20)	18 (9)
Age 30–59	27 (16)	30 (16)	15 (9)
Age 60–79	18 (11)	16 (7)	11 (6)
Males			
BMI < 24			
Age 19–29	37 (19)	46 (34)	23 (14)
Age 30–59	24 (9)	23 (19)	17 (10)
Age 60–79	16 (4)	21 (5)	10 (2)
BMI ≥ 24			
Age 19–29	29 (10)	27 (15)	17 (8)
Age 30–59	15 (6)	16 (9)	10 (6)
Age 60–79	13 (7)	10 (5)	7 (5)
All subjects	27 (17)	29 (22)	16 (11)

MEDIAN AND ULNAR SENSORY STUDIES TO THE 4TH DIGIT

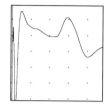

Typical waveform appearance

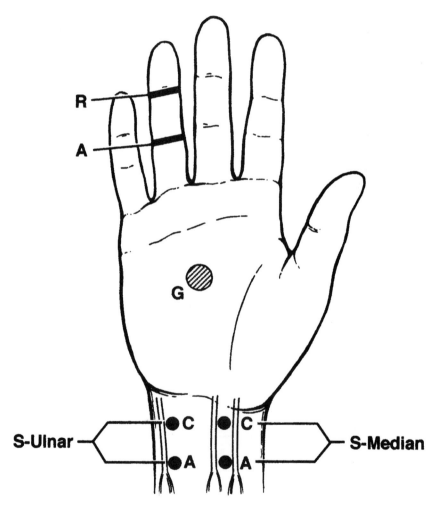

Electrode Placement

Active electrode (A): A ring electrode is placed on the proximal interphalangeal joint of the 4th digit.

Reference electrode (R): A ring electrode is placed on the distal interphalangeal joint of the 4th digit.

Ground electrode (G): Placement is on the dorsum of the hand.

Median stimulation point (s-median): Wrist stimulation is accomplished with the cathode (C) 14 cm proximal to the active electrode, slightly ulnar to the tendon of the flexor carpi radialis. The anode (A) is proximal.

Ulnar stimulation point (s-ulnar): Wrist stimulation is accomplished with the cathode (C) 14 cm proximal to the active electrode, slightly radial to the tendon of the flexor carpi ulnaris. The anode (A) is proximal.

Machine settings: Sensitivity—20 µV/division, Low frequency filter—2 Hz, High frequency filter—10 kHz, Sweep speed—1 msec/division.

Nerve fibers tested: Median: C8 nerve root through the lower trunk, anterior division, and medial cord of the brachial plexus. Ulnar: C8 nerve root through the lower trunk, anterior division, and medial cord of the brachial plexus.

Normal values (1) (30 subjects) (temperature over wrist and finger at least 31 degrees Celsius):

Onset latencies (msec)

	Mean	S.D.	Mean + 2 S.D.
Median	2.4	0.2	2.8
Ulnar	2.3	0.2	2.7

Peak latencies (msec)

	Mean	S.D.	Mean + 2 S.D.
Median	2.9	0.2	3.3
Ulnar	2.9	0.2	3.3

Onset to peak amplitude (μV)

	Mean	S.D.
Median	29.1	10.6
Ulnar	24.9	10.1

Peak to peak amplitude (μV)

	Mean	S.D.
Median	38.2	15.0
Ulnar	30.9	14.6

Duration (onset to return to baseline): (msec)

	Mean	S.D.
Median	1.0	0.1
Ulnar	1.1	0.2

Duration (onset to positive peak): (msec)

	Mean	S.D.
Median	1.3	0.2
Ulnar	1.4	0.4

Acceptable Differences

The upper limit of normal increase in latency for the median versus ulnar nerve is 0.3–0.5 msec (2–5).

The upper limit of normal side to side difference in onset and peak latency for both the median and ulnar studies is 0.5 msec (1).

The upper limit of normal side to side difference in duration is 0.5 msec (1).

Helpful Hints

- Anomalous innervation patterns may be encountered.
- To avoid contamination of the response by contact of the ring electrodes with the adjacent fingers, a small roll of gauze may need to be placed between the digits to hold them apart.

Notes _____

_____

_____

_____

_____

_____

_____

_____

_____

REFERENCES

1. DiBenedetto M, Mitz M, Klingbeil GE, Davidoff D: New criteria for sensory nerve conduction especially useful in diagnosing carpal tunnel syndrome. *Arch Phys Med Rehabil* 1986; 67:586–589.
2. Uncini A, Lange DJ, Solomon M, et al: Ring finger testing in carpal tunnel syndrome: a comparative study of diagnostic utility. *Muscle Nerve* 1989; 12:735–741.
3. Jackson DA, Clifford JC: Electrodiagnosis of mild carpal tunnel syndrome. *Arch Phys Med Rehabil* 1989; 70:199–204.

4. Johnson EW, Kukla RD, Wongsam PE, et al: Sensory latencies to the ring finger: normal values and relation to carpal tunnel syndrome. *Arch Phys Med Rehabil* 1981; 62:206–208.
5. Cho DS, MacLean DC: Comparison of normal values of median, radial, and ulnar sensory latencies. *Muscle Nerve* 1984; 7:575.

Chapter 3

Lower Extremity Motor Nerves

FEMORAL MOTOR NERVE TO THE QUADRICEPS

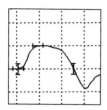

Typical waveform appearance

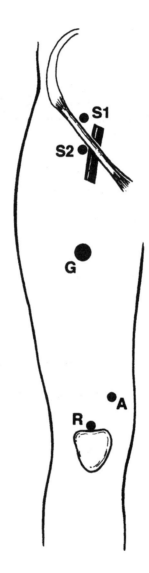

Electrode Placement

Active electrode (A): Placement is over the center of the vastus medialis.

Reference electrode (R): Placement is over the quadriceps tendon just proximal to the patella (1). Placement over the patella has also been described (2).

Ground electrode (G): Placement is between the stimulating and recording electrodes.

Stimulation point 1 (S1): A monopolar needle electrode is used as the cathode for all but the most thin subjects (2). The stimulus is applied superior to the inguinal ligament just lateral to the femoral artery. The anode (A) is under the buttock.

Stimulation point 2 (S2): A needle electrode is used as well, but inferior to the inguinal ligament and lateral to the femoral artery.

Machine settings: Sensitivity—1 mV/division, Sweep speed—2 msec/division.

Nerve fibers tested: L2, L3, and L4 nerve roots, through the posterior division of the lumbosacral plexus.

Normal values (1) (100 subjects):

Latency for stimulation above ligament (msec)

Mean	S.D	Mean + 2 S.D.	Range
7.1	0.7	8.5	6.1–8.4

Latency for stimulation below ligament (msec)

Mean	S.D.	Mean + 2 S.D.	Range
6.0	0.7	7.4	5.5–7.5

Delay across inguinal ligament (msec)

Mean	S.D.	Mean + 2 S.D.	Range
1.1	0.4	1.9	0.8–1.8

Amplitude (mV) (2)

Range
0.2–11.0

Helpful Hints

- Too low a stimulus intensity may result in an H-reflex being elicited.
- In the cited reference, only 75 of 100 studies were included in the final results because satisfactory recordings were not obtained in all subjects. Certain portions of the study for the remaining 25 studies were also used. The subjects were not true "normals." The data were derived using surface stimulation.
- Proper stimulator placement can be confirmed by observing contraction of quadriceps.
- The length of the femoral nerve segment in the cited reference was 35.4 ± 1.9 cm (range 29–38).
- The distance between stimulation point above and below the inguinal ligament in the cited reference was 5.5 ± 1.6 cm (range 4.2–6.6).

Notes

REFERENCES

1. Johnson EW, Wood PK, Powers JJ: Femoral nerve conduction studies. *Arch Phys Med Rehabil* 1968; 49:528–532.
2. Kraft GH, Johnson EW: Proximal motor nerve conduction and late responses: an AAEM workshop. American Association of Electrodiagnostic Medicine, Rochester, Minnesota, 1986.

ADDITIONAL READING/ALTERNATE TECHNIQUE

1. Gassel MM: A study of femoral nerve conduction time. *Arch Neurol* 1963; 9:607–614.

PERONEAL NERVE

PERONEAL MOTOR NERVE TO THE EXTENSOR DIGITORUM BREVIS

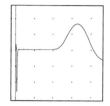

Typical waveform appearance

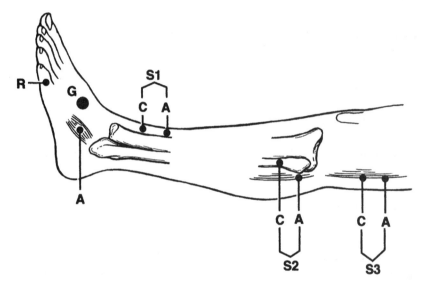

Electrode Placement

Active electrode (A): Placement is over the midpoint of the extensor digitorum brevis muscle on the dorsum of the foot.

Reference electrode (R): Placement is slightly distal to the 5th metatarsophalangeal joint.

Ground electrode (G): Placement is over the dorsum of the foot.

Stimulation point 1 (S1): The cathode (C) is placed 8 cm proximal to the active electrode, slightly lateral to the tibialis anterior tendon. The anode (A) is proximal.

Stimulation point 2 (S2): The cathode (C) is slightly posterior and inferior to the fibular head. The anode (A) is proximal.

Stimulation point 3 (S3): The cathode (C) is placed approximately 10 cm proximal to the S2 cathode placement and medial to the tendon of the biceps femoris. The anode (A) is proximal.

F-wave stimulation: The cathode is positioned as for S1, but with the anode distally.

Machine settings: Sensitivity- 5 mV/division, Low frequency filter—2–3 Hz, High frequency filter—10 kHz, Sweep speed—5 msec/division.

Nerve fibers tested: L5 and S1 nerve roots, through the posterior division of the lumbosacral plexus, and the sciatic and common peroneal nerves.

Normal values (1) (242 subjects) (skin temperature over the dorsum of the foot greater than or equal to 31 degrees Celsius):

Onset latency (msec)

Mean	S.D.	Mean + 2 S.D.	97th Percentile
4.8	0.8	6.4	6.5

Amplitude (mV)

Age Range	Mean	S.D.	3rd Percentile
19–39	6.8	2.5	2.6
40–79	5.1	2.5	1.1
All subjects	5.9	2.6	1.3

Area (μVsec)

Age Range	Mean	S.D.	3rd Percentile
19–49	20.2	8.0	6.8
50–79	14.9	7.6	3.6
All subjects	18.3	8.2	4.3

Duration (msec)

Mean	S.D.	Mean + 2 S.D.	97th Percentile
5.7	1.0	7.7	7.7

Nerve conduction velocity (m/sec)

		Mean	S.D.	Mean – 2 S.D.	3rd Percentile
S1–S2					
Height in cm (in.)	Age				
<170 cm (<5′7″)	19–39	49	4	41	43
	40–79	47	5	37	39
≥ 170 cm (≥ 5′7″)	19–39	46	4	38	37
	40–79	44	4	36	36
All subjects		47	4	39	38
S2–S3					
All subjects		57	9	39	42

F-wave latencies (2) (180 subjects)—shortest of 10 stimuli

Age 19–39

Height in cm (in.)	Mean	S.D.	Mean + 2 S.D.	97th Percentile
<160 (<5'3")	43.6	2.5	48.6	
160–169 (5'3"–5'6")	47.1	3.7	54.5	
≥ 170 (≥ 5'7")	51.5	4.1	59.7	

Age 40–79

Height in cm (in.)	Mean	S.D.	Mean + 2 S.D.	97th Percentile
<160 (<5'3")	45.4	4.8	55.0	
160–169 (5'3"–5'6")	49.6	4.6	58.8	
≥ 170 (≥5'7")	54.6	4.5	63.6	
All subjects	50.2	5.5	61.2	60.5

Acceptable Differences

The upper limit of normal increase in latency from one side to the other is 1.6 msec.

The upper limit of normal decrease in amplitude from one side to the other is 61%.

The upper limit of normal decrease in S1–S2 nerve conduction velocity from one side to the other is 8 m/sec.

The upper limit of normal decrease in S2–S3 nerve conduction velocity from one side to the other is 19 m/sec.

The upper limit of normal decrease in nerve conduction velocity from S1–S2 to S2–S3 is 6 m/sec.

The upper limit of normal decrease in amplitude from S1 to S2 is 32%.

The upper limit of normal decrease in amplitude from S2 to S3 is 25%.

The upper limit of normal side to side difference in the shortest F-wave latency is 5.1 msec.

Helpful Hints

- Care must be taken at popliteal stimulation to not concomitantly activate the tibial nerve.

- An accessory peroneal nerve is commonly present (20–25% incidence), although it is less commonly of clinical significance (3,4). The accessory peroneal nerve passes behind the lateral malleolus to innervate the extensor digitorum brevis. Its presence should be suspected if the amplitude to proximal stimulation is greater than on ankle stimulation. Its presence can be confirmed by stimulating behind the lateral malleolus. If a response is recorded from the extensor digitorum brevis, an accessory peroneal nerve is present.

- A short segment incremental stimulation ("inching") technique has been described for testing the conduction of the peroneal nerve across the knee. The nerve is stimulated in 2 cm increments starting 4 cm distal and proceeding to 6 cm proximal to the head of the fibula. In normal subjects the difference in latency between successive stimulation points varies from 0.2 to 0.65 msec. Abrupt waveform changes or decreases in amplitude may be a sign of conduction block (5).

Notes

REFERENCES

1. Buschbacher RM: Peroneal nerve motor conduction to the extensor digitorum brevis. *Am J Phys Med Rehabil* 1999; 78:S26–S31.
2. Buschbacher RM: Peroneal F-waves recorded from the extensor digitorum brevis. *Am J Phys Med Rehabil* 1999; 78:S48–S52.
3. Crutchfield CA, Gutmann L: Hereditary aspects of accessory deep peroneal nerve. *J Neurol Neurosurg Psychiatry* 1973; 36:989–990.
4. Neundoerfer B, Seiberth R: The accessory deep peroneal nerve. *J Neurol* 1975; 209:125–129.
5. Kanakamedala RV, Hong CZ: Peroneal nerve entrapment at the knee localized by short segment stimulation. *Am J Phys Med Rehabil* 1989; 68:116–122.

ADDITIONAL READINGS/ALTERNATE TECHNIQUES

1. Checkles NS, Bailey JA, Johnson EW: Tape and caliper surface measurements in determination of peroneal nerve conduction velocity. *Arch Phys Med Rehabil* 1969; 50:214–218.
2. Jimenez J, Easton JKM, Redford JB: Conduction studies of the anterior and posterior tibial nerves. *Arch Phys Med Rehabil* 1970; 51:164–169.
3. Falco FJE, Hennessey WJ, Goldberg G, Braddom RL: Standardized nerve conduction studies in the lower limb of the healthy elderly. *Am J Phys Med Rehabil* 1994; 73:168–174.

PERONEAL MOTOR NERVE TO THE PERONEUS BREVIS

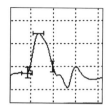

Typical waveform appearance

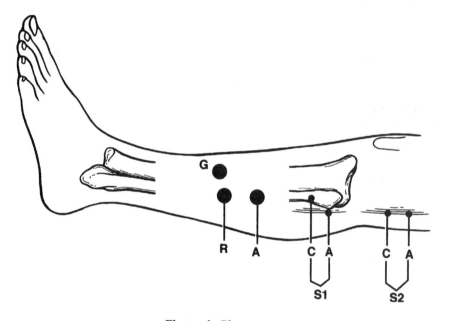

Electrode Placement

Active electrode (A): A 32 mm disc electrode is placed two fifths of the distance from the head of the fibula to the tip of the lateral malleolus.

Reference electrode (R): Placement is 4 cm distal to the active electrode over the muscle tendon.

Ground electrode (G): Placement is over the tibia, 3–4 cm distal to the reference electrode.

Stimulation point 1 (S1): The cathode is placed slightly below the head of the fibula. The anode is proximal.

Stimulation point 2 (S2): The cathode is placed just medial to the lateral border of the popliteal space at the level of the mid-patella, approximately 10 cm proximal to stimulation point 1.

Machine settings: Standard motor settings are used.

Nerve fibers tested: L5, S1, and S2 nerve roots, through the posterior division of the sacral plexus and the sciatic nerve.

Normal values (1) (34 subjects) (room temperature 22.2–23.3 degrees Celsius):

Onset latency (msec)

Mean	S.D.	Mean + 2 S.D.	Range
3.0	0.8	4.6	1.7–5.4

Amplitude (mV)

Mean	S.D.
5.3	1.7

Nerve conduction velocity (m/sec)

Mean	S.D.	Mean − 2 S.D.
55.3	10.2	34.9

Notes

REFERENCE

1. Devi S, Lovelace RE, Duarte N: Proximal peroneal nerve conduction velocity: recording from anterior tibial and peroneus brevis muscles. *Ann Neurol* 1977; 2:116–11.

PERONEAL MOTOR NERVE TO THE PERONEUS LONGUS

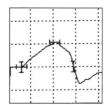

Typical waveform appearance

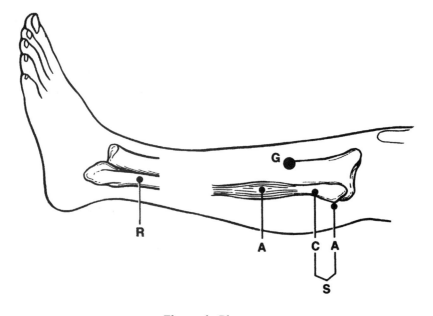

Electrode Placement

Active electrode (A): Placement is 8 cm from the cathode over the peroneus longus on the lateral surface of the fibula.

Reference electrode (R): Placement is at the ankle over the tendon of the peroneus longus muscle.

Ground (G): Placement is over the upper anterior lower leg.

Stimulation point (S): The cathode (C) is placed at the posterolateral aspect of the fibular neck. The anode (A) is proximal.

Machine settings: Sensitivity—2 mV/division (1 mV to determine onset latency), Low frequency filter—2 Hz, High frequency filter—10 kHz, Sweep speed—2 msec/division.

Nerve fibers tested: L5, S1, and S2 nerve roots, through the posterior division of the lumbosacral plexus and the sciatic nerve.

Normal values (1) (81 subjects) (skin temperature over the lateral surface just below the knee joint greater than or equal to 32 degrees Celsius):

Onset latency (msec)

Mean	S.D.	Mean + 2 S.D.	Range
2.6	0.2	3.0	1.9–3.0

Amplitude (mV)

Mean	S.D.	Range
6.2	1.7	3.4–10.6

Helpful Hints

- This technique utilizes a fixed distance measurement from a given stimulation site rather than from a given recording site. Therefore, the active electrode is not always over the motor point. This

(cont.)

may cause submaximal amplitude measurements to be recorded. It is also not known how accurate the latency measurements are with such a technique. Caution is advised when using such a technique for persons at the extremes of height.

- The recorded action potential may exhibit multiple peaks, possibly from the volume conducted potentials of adjacent muscles.

Notes

REFERENCE

1. Lee HJ, Bach JR, DeLisa JA: Peroneal nerve motor conduction to the proximal muscles: an alternative approach to conventional methods. *Am J Phys Med Rehabil* 1997; 76:197–199.

PERONEAL MOTOR STUDY TO THE TIBIALIS ANTERIOR

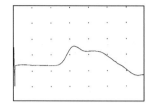

Typical waveform appearance

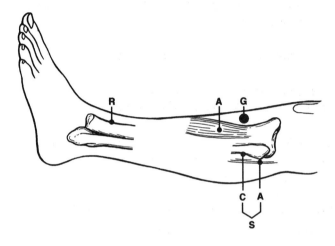

Fixed Distance Technique

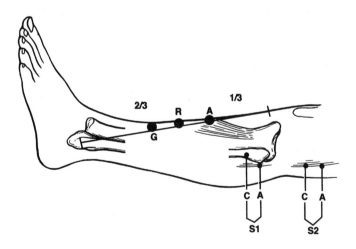

Motor Point Recording Technique

Electrode Placement

Fixed Distance Technique (1)

Active electrode (A): Placement is 8 cm from the cathode, in approximately a 45 degree angle from the cathode, over the tibialis anterior muscle.

Reference electrode (R): Placement is at the ankle over the tendon of the tibialis anterior muscle.

Ground (G): Placement is over the upper anterior lower leg.

Stimulation point (S): The cathode (C) is placed at the posterolateral aspect of the fibular neck. The anode (A) is proximal.

Machine settings: Sensitivity—2 mV/division (1 mV to determine onset latency), Low frequency filter—2 Hz, High frequency filter—10 kHz, Sweep speed—2 msec/division.

Nerve fibers tested: L4 and L5 nerve roots, through the posterior division of the lumbosacral plexus and the sciatic nerve.

Normal values (1) (81 subjects) (skin temperature over the lateral surface just below the knee joint greater than or equal to 32 degrees Celsius):

Onset latency (msec)

Mean	S.D.	Mean + 2 S.D.	Range
2.5	0.3	3.1	2.0–3.0

Amplitude (mV)

Mean	S.D.	Range
6.2	1.3	3.6–9.3

Motor Point Recording Technique (2)

Active electrode (A): A 32 mm disc electrode is placed one third of the distance from the tibial tuberosity to the tip of the lateral malleolus.

Reference electrode (R): Placement is over the medial tibia, 4 cm distal to the active electrode.

Ground electrode (G): Placement is on the tibia, 3–4 cm distal from the reference electrode.

Stimulation point 1 (S1): The cathode (C) is placed slightly below the the head of the fibula. The anode (A) is proximal.

Stimulation point 2 (S2): The cathode (C) is placed just medial to the lateral border of the popliteal space at the level of the mid-patella, approximately 10 cm proximal to stimulation point 1.

Nerve fibers tested: L4 and L5 nerve roots, through the posterior division of the sacral plexus and the sciatic and common peroneal nerves.

Normal values (1) (34 subjects) (room temperature 22.2–23.3 degrees Celsius):

Onset latency (msec)

Mean	S.D.	Mean + 2 S.D.	Range
3.0	0.6	4.2	2.0–4.4

Amplitude (mV)

Mean	S.D.
3.9	1.2

Nerve conduction velocity (m/sec)

Mean	S.D.	Mean – 2 S.D.
66.3	12.9	40.5

Helpful Hints

- The first technique utilizes a fixed distance measurement from a given stimulation site rather than from a given recording site. Therefore, the active electrode is not always over the motor point. This may cause submaximal amplitude measurements to be recorded. It is also not known how accurate the latency measurements are with such a technique. Caution is advised when using such a technique for persons at the extremes of height or calf circumference.

- With the first technique, the recorded action potential may exhibit multiple peaks, possibly from the volume conducted potentials of adjacent muscles.

Notes _____

_____
_____
_____
_____
_____
_____
_____

REFERENCES

1. Lee HJ, Bach JR, DeLisa JA: Peroneal nerve motor conduction to the proximal muscles: an alternative approach to conventional methods. *Am J Phys Med Rehabil* 1997; 76:197–199.
2. Devi S, Lovelace RE, Duarte N: Proximal peroneal nerve conduction velocity: recording from anterior tibial and peroneus brevis muscles. *Ann Neurol* 1977; 2:116–119.

SCIATIC NERVE

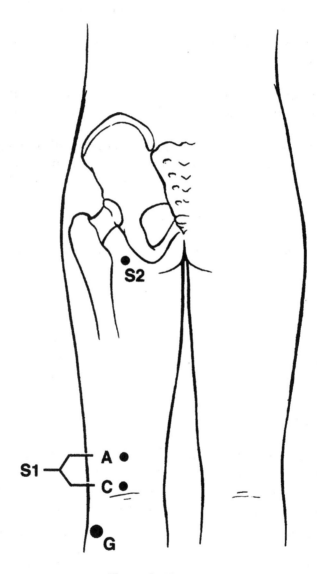

Electrode Placement

Recording electrodes: Placement is on the distal muscles of the foot, such as the extensor digitorum brevis (peroneal portion) abductor hallucis (tibial portion), or abductor digiti minimi (tibial portion) (see other sections of this book describing recording from these sites) (1,2).

Ground electrode (G): Placement is between the stimulating and recording electrodes.

Stimulation point 1 (S1): Surface stimulation is applied in the popliteal fossa with the cathode (C) distal and the anode (A) proximal.

Stimulation point 2 (S2): A long needle electrode (cathode) is used to stimulate the sciatic nerve just below the gluteal fold in a line directly above the apex of the popliteal fossa. The anode is placed nearby.

Machine settings: Standard motor settings are used.

Normal values (1) (18 subjects) (room temperature 23–26 degrees Celsius):

Nerve conduction velocity (m/sec)

	Mean	S.D.	Range
Tibial portion	52.75	4.66	46.7–59.6
Peroneal portion	54.33	4.36	48.5–61.5

Helpful Hints

- Finding the correct gluteal fold stimulation site may be difficult. The peroneal portion of the nerve lies a bit lateral, with the tibial portion being more medial. Observing the foot motion will help identify which portion is being stimulated the most (1).
- With proximal stimulation the recordings made at the extensor digitorum brevis may include volume conducted potentials from other foot muscles that are innervated by the tibial portion of the nerve. (2)
- Stimulation can also be applied at the ankle to calculate conduction velocity along the lower leg.

Notes

REFERENCES

1. Ma DM, Liveson JA: *Nerve conduction handbook.* Philadelphia: FA Davis, 1983.
2. Yap CB, Hirota T: Sciatic nerve motor conduction velocity study. *J Neurol Neurosurg Psychiatry* 1967; 30:233–239.

ADDITIONAL READINGS/ALTERNATE TECHNIQUES

1. Gassel MM, Trojaborg W: Clinical and electrophysiological study of the pattern of conduction times in the distribution of the sciatic nerve. *J Neurol Neurosurg Psychiatry* 1964; 27:351–357.
2. Inaba A, Yokota T, Komori T, Hirose K: Proximal and segmental motor nerve conduction in the sciatic nerve produced by percutaneous high voltage electrical stimulation. *Electroencephalogr clin Neurophysiol* 1996; 101:100–104.

TIBIAL NERVE

TIBIAL MOTOR NERVE (INFERIOR CALCANEAL BRANCH) TO THE ABDUCTOR DIGITI MINIMI

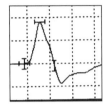

Typical waveform appearance

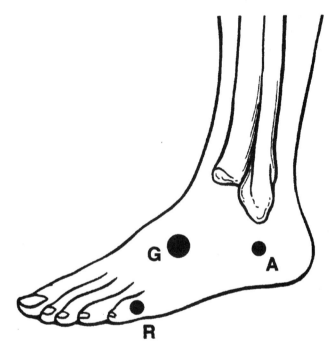

Electrode Placement

Active electrode (A): Placement is directly below the lateral malleolus, halfway between the tip of the malleolus and the sole of the foot.

Reference electrode (R): Placement is on the 5th digit.

Ground electrode (G): Placement is on the dorsum of the foot.

Stimulation point: Stimulation is similar to that used for the medial plantar branch to the abductor hallucis. In the cited study, the medial recording site was 1 cm inferior and posterior to the navicular tubercle. The stimulation was applied 8 cm and 10 cm proximal (measured along the course of the nerve with the ankle in neutral) to the recording site. These same stimulation sites are used to test the lateral site. The cathode is distal and the anode proximal.

Machine settings: Sensitivity—500 µV/division, Low frequency filter—8 Hz, High frequency filter—8 kHz, Sweep speed—5 msec/division (1).

Nerve fibers tested: S1 and S2 nerve roots, through the anterior division of the lumbosacral plexus and the sciatic nerve. Recent research indicates that the abductor digiti minimi is innervated by the inferior calcaneal branch of the tibial nerve (2).

Normal values (3) (37 subjects) (skin temperature below the medial malleolus 29–34 degrees Celsius) (1):

Onset latency (msec)

	Mean	S.D.	Mean + 2 S.D.
Medial recording (abductor hallucis)			
8 cm distance	3.4	0.5	4.4
10 cm distance	3.8	0.5	4.8
Lateral recording (abductor digiti minimi)			
8 cm distance	3.6	0.5	4.6
10 cm distance	3.9	0.5	4.9

Notes

REFERENCES

1. DeLisa JA, Lee HJ, Baran EM, et al: *Manual of nerve conduction velocity and clinical neurophysiology*, 3rd ed. New York: Raven Press, 1994.
2. Del Toro DR, Mazur A, Dwzierzynski WW, Park TA: Electrophysiologic mapping and cadaveric dissection of the lateral floor: implications for tibial motor nerve conduction studies. *Arch Phys Med Rehabil* 1998; 78:823–826.
3. Fu R, DeLisa JA, Kraft GH: Motor nerve latencies through the tarsal tunnel in normal adult subjects: standard determinations corrected for temperature and distance. *Arch Phys Med Rehabil* 1980; 61:243–248.

ADDITIONAL READINGS/ALTERNATE TECHNIQUES

1. Oh SJ, Sarala PK, Kuba T, Elmore RS: Tarsal tunnel syndrome: electrophysiological study. *Ann Neurol* 1979; 5:327–330.
2. Irani KD, Grabois M, Harvey SC: Standardized technique for diagnosis of tarsal tunnel syndrome. *Am J Phys Med Rehabil* 1982; 61:26–31.

TIBIAL MOTOR NERVE (MEDIAL PLANTAR BRANCH) TO THE ABDUCTOR HALLUCIS

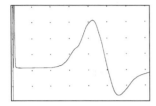

Typical waveform appearance

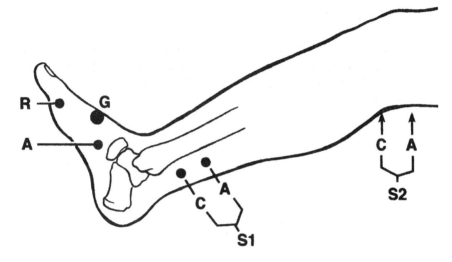

Electrode Placement

Active electrode (A): Placement is over the medial foot, slightly anterior and inferior to the navicular tubercle (at the most superior point of the arch formed by the junction of plantar skin and dorsal foot skin).

Reference electrode (R): Placement is slightly distal to the 1st metatarsophalangeal joint, on the medial surface of the joint.

Ground electrode (G): Placement is over the dorsum of the foot.

Stimulation point 1 (S1): The cathode (C) is placed 8 cm proximal to the active electrode (measured in a straight line with the ankle in neutral position) and slightly posterior to the medial malleolus. The anode (A) is proximal.

Stimulation point 2 (S2): The cathode (C) is placed at the mid-popliteal fossa or slightly medial or lateral to the midline. The anode (A) is proximal.

F-wave stimulation: The cathode is positioned as for stimulation point 1, but with the anode distally.

Machine settings: Sensitivity—5 mV/division, Low frequency filter—2–3 Hz, High frequency filter—10 kHz, Sweep speed—5 msec/division

Nerve fibers tested: S1 and S2 nerve roots, through the anterior division of the lumbosacral plexus and the sciatic nerve.

Normal values (1) (250 subjects) (skin temperature over the dorsum of the foot greater than or equal to 31 degrees Celsius):

Onset latency (msec)

Mean	S.D.	Mean + 2 S.D.	97th Percentile
4.5	0.8	6.1	6.1

Amplitude (mV)

Age Range	Mean	S.D.	Mean − 2 S.D.	3rd Percentile
19–29	15.3	4.5	6.3	5.8
30–59	12.9	4.5	3.9	5.3
60–79	9.8	4.2	1.4	1.1
All subjects	12.9	4.8	3.3	4.4

Area (µVsec)

Age Range	Mean	S.D.	Mean − 2 S.D.	3rd Percentile
19–49	38.9	14.3	10.3	14.2
50–79	29.2	13.3	2.6	5.0
All subjects	35.3	14.7	5.9	9.1

Duration (msec)

Mean	S.D.	Mean + 2 S.D.	97th Percentile
6.3	1.2	8.7	8.7

Nerve conduction velocity (m/sec)

Age 19–49

Height in cm (in.)	Mean	S.D.	Mean − 2 S.D.	3rd Percentile
<160 (<5′3″)	51	4	43	44
160–169 (5′3″–5′6″)	49	6	37	42
≥ 170 (≥ 5′7″)	47	5	37	37

Age 50–79

Height in cm (in.)	Mean	S.D.	Mean − 2 S.D.	3rd Percentile
<160 (<5′3″)	49	5	39	40
160–169 (5′3″–5′6″)	45	5	35	37
≥ 170 (≥ 5′7″)	44	5	34	34
All subjects	47	6	35	39

F-wave latencies (msec) (2) (180 subjects)—shortest of 10 stimuli

Age 19–39

Height in cm (in.)	Mean	S.D.	Mean + 2 S.D.	97th Percentile
<160 (<5'3")	43.2	2.2	47.6	
160–169 (5'3"–5'6")	47.2	3.0	53.2	
170–179 (5'7"–5'10")	52.0	4.0	60.0	
≥ 180 (≥5'11")	53.1	4.4	61.9	

Age 40–59

Height in cm (in.)				
<160 (<5'3")	45.4	4.0	53.4	
160–169 (5'3"–5'6")	49.3	2.2	53.7	
170–179 (5'7"–5'10")	53.6	3.7	61.0	
<180 (≥5'11")	58.3	5.3	68.9	

Age 60–79

Height in cm (in.)				
<160 (<5'3")	49.0	4.8	58.6	
160-169 (5'3"–5'6")	52.8	4.4	61.6	
170-179 (5'7"–5'10")	54.7	3.2	61.1	
≥ 180 (≥5'11")	57.3	5.8	68.9	
All subjects	50.8	5.3	61.4	59.8

Acceptable Differences

The upper limit of normal increase in latency from one side to the other is 1.4 msec.

The upper limit of normal decrease in amplitude from one side to the other is 50%.

The upper limit of normal decrease in nerve conduction velocity from one side to the other is 10 m/sec.

The upper limit of normal decrease in amplitude from ankle to knee stimulation is 71%.

The upper limit of normal side to side difference in the shortest F-wave latency is 5.7 msec.

Helpful Hints

- Ankle stimulation should be approximately halfway between the medial malleolus and the Achilles tendon.

- Care should be taken to not concomitantly stimulate the peroneal nerve at the knee. Stimulation should be close to the midline of the popliteal fossa, but the stimulator may need to be moved slightly medially or laterally to obtain an optimal response. Watching for the direction of foot motion on stimulation will help ensure that the proper nerve is stimulated.

- Obtaining a true supramaximal response at knee stimulation may be difficult at times. The amplitude drop with knee versus ankle stimulation for this nerve study is greater than that seen with most other nerve studies.

- In subjects with large feet, the 8 cm fixed distance between stimulating and recording electrodes may fail to include the entire "tarsal tunnel" area. A 10 cm distance can be used (3,4). With the active electrode 1 cm posterior and inferior to the navicular tubercle and a 10 cm distance between the stimulating and recording electrodes, the mean latency has been described as 3.8 ± 0.5 msec (at 29–34 degrees Celsius) (3). In a series of elderly subjects using slightly different methodology from the data presented above, the mean latency at 10 cm (again recording posterior and inferior to the navicular tubercle) was 4.5 ± 0.7 msec. (4)

REFERENCES

1. Buschbacher RM: Tibial nerve motor conduction to the abductor hallucis. *Am J Phys Med Rehabil* 1999; 78:S15–S20.
2. Buschbacher RM: Tibial nerve F-waves recorded from the abductor hallucis. *Am J Phys Rehabil* 1999; 78:S42–S47.
3. Fu R, DeLisa JA, Kraft GH: Motor nerve latencies through the tarsal tunnel in normal adult subjects: standard determinations corrected for temperature and distance. *Arch Phys Med Rehabil* 1980; 61:243–248.
4. Falco FJE, Hennessey WJ, Goldberg G, Braddom RL: Standardized nerve conduction studies in the lower limb of the healthy elderly. *Am J Phys Med Rehabil* 1994; 73:168–174.

ADDITIONAL READINGS/ALTERNATE TECHNIQUES

1. Jimenez J, Easton JKM, Redford JB: Conduction studies of the anterior and posterior tibial nerves. *Arch Phys Med Rehabil* 1970; 51:164–169.
2. Oh SJ, Sarala PK, Kuba T, Elmore RS: Tarsal tunnel syndrome: electrophysiological study. *Ann Neurol* 1979; 5:327–330.
3. Irani KD, Grabois M, Harvey SC: Standardized technique for diagnosis of tarsal tunnel syndrome. *Am J Phys Med Rehabil* 1982; 61:26–31.
4. Felsenthal G, Bulter DH, Shear MS: Across-tarsal-tunnel motor-nerve conduction technique. *Arch Phys Med Rehabil* 1992; 73:64–69.

TIBIAL MOTOR NERVE (LATERAL PLANTAR BRANCH) TO THE FLEXOR DIGITI MINIMI BREVIS

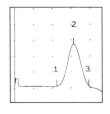

Typical waveform appearance

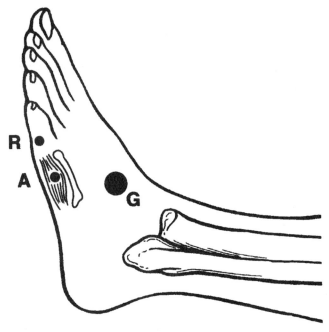

Electrode Placement

Active electrode (A): Placement is on the midpoint of the inferolateral edge of the 5th metatarsal.

Reference electrode (R): Placement is slightly distal to the 5th metatarsophalangeal joint on the lateral surface of the joint.

Ground electrode (G): Placement is on the dorsum of the foot.

Stimulation point (S): Stimulation is applied at the same point as for the study recording from the abductor hallucis (see section on the tibial motor nerve study to the abductor hallucis). The cathode (C) is placed behind the medial malleolus 8 cm proximal to a point slightly anterior and inferior to the navicular tubercle on the medial foot. The anode (A) is proximal.

Machine settings: Sensitivity—5 mV/division, Low frequency filter—2–3 Hz, High frequency filter—10 kHz, Sweep speed—5 msec/division.

Nerve fibers tested: S1 and S2 nerve roots, through the anterior division of the lumbosacral plexus and the sciatic nerve.

Normal values (1) (205 subjects) (skin temperature over the dorsum of the foot at least 31 degrees Celsius):

Onset latency (msec)

Mean	S.D.	Mean + 2 S.D.	97th Percentile
6.4	1.0	8.4	8.3

Amplitude (mV)

Age Range	Mean	S.D.	3rd Percentile
19–29	7.8	3.2	2.8
30–59	6.0	3.2	1.7
60–79	4.7	3.1	1.0
All subjects	6.1	3.3	1.4

Area (µVsec)

Age Range	Mean	S.D.	3rd Percentile
19–59	17.5	9.5	4.2
60–79	11.8	6.6	3.1
All subjects	16.2	9.2	3.1

Duration (msec)

Mean	S.D.	97th Percentile
5.8	1.7	10.0

Acceptable Differences

The upper limit of normal increase in latency from one side to the other is 1.5 msec.

The upper limit of normal decrease in amplitude from one side to the other is 58%.

Normally the lateral branch latency is greater than the medial branch latency. The upper limit of normal increase in latency to the flexor digiti minimi brevis versus the abductor hallucis is 3.5 msec. If the medial latency comes within 0.3 msec of the lateral latency or exceeds the lateral latency, this is a sign of medial branch slowing.

Helpful Hints

- This study is very easy to perform and allows comparison of the lateral tibial branch versus the medial branch.
- Previous authors have used this same recording site but have remarked that the recording is from the abductor digiti minimi. Recent research has identified the source of this potential to be coming from the flexor digiti minimi brevis (2).

Notes

REFERENCES

1. Buschbacher RM. Tibial motor conduction to the flexor digiti minimi brevis. *Am J Phys Med Rehabil* 1999; 78:S21–S25.
2. Del Toro DR, Mazur A, Dwzierzynski WW, Park TA: Electrophysiologic mapping and cadaveric dissection of the lateral floor: implications for tibial motor nerve conduction studies. *Arch Phys Med Rehabil* 1998; 78:823–826.

ADDITIONAL READINGS/ALTERNATE TECHNIQUES

1. Oh SJ, Sarala PK, Kuba T, Elmore RS: Tarsal tunnel syndrome: electrophysiological study. *Ann Neurol* 1979; 5:327–330.
2. Fu R, DeLisa JA, Kraft GH: Motor nerve latencies through the tarsal tunnel in normal adult subjects: standard determinations corrected for temperature and distance. *Arch Phys Med Rehabil* 1980; 61:243–248.
3. Irani KD, Grabois M, Harvey SC: Standardized technique for diagnosis of tarsal tunnel syndrome. *Am J Phys Med Rehabil* 1982; 61:26–31.
4. Felsenthal G, Bulter DH, Shear MS: Across-tarsal-tunnel motor-nerve conduction technique. *Arch Phys Med Rehabil* 1992; 73:64–69.

H-REFLEX TO THE CALF

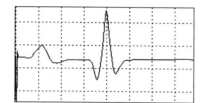

Typical waveform appearance

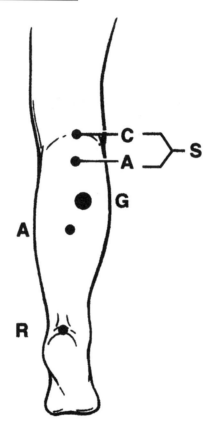

Electrode Placement

Active electrode (A): The subject is placed prone. The knee is passively flexed by the examiner to mark the popliteal crease. The lower leg is then lowered onto a pillow with the foot hanging over the edge of the table and the ankle slightly plantarflexed. A second mark is placed on the posterior calcaneus. The distance between the two marked points is measured and the active electrode is placed at the midpoint.

Reference electrode (R): Placement is over the posterior calcaneus.

Ground electrode (G): Placement is between the stimulating and recording electrodes.

Stimulation point (S): The cathode (C) is placed at the mid-popliteal crease with the anode (A) distal.

Machine settings: Sensitivity—500 µV/division, Low frequency filter—2–3 Hz, High frequency filter—10 kHz, Sweep speed—10 msec/division, Stimulus duration—1.0 msec

Nerve fibers tested: S1 nerve roots, afferent and efferent sciatic nerve, and their spinal cord connection.

(cont.)

Normal values (1) (251 subjects): (skin temperature over the ankle 31 degrees Celsius or greater):

Onset latency (msec)

Age 19–39

Height in cm (in.)	Mean	S.D.	Mean + 2 S.D.	97th Percentile
<160 (<5′3″)	27.1	1.8	30.7	29.8
160–169 (5′3″–5′6″)	28.6	1.9	32.4	32.3
170–179 (5′7″–5′10″)	30.3	1.8	33.9	33.7
≥ 180 (≥5′11″)	32.0	2.1	36.2	35.9

Age 40–49

Height in cm (in.)

<160 (<5′3″)	27.8	1.1	30.0	29.8
160–169 (5′3″–5′6″)	30.2	1.4	33.0	32.8
170–179 (5′7″–5′10″)	31.0	1.6	34.2	33.7
≥ 180 (≥5′11″)	32.7	2.1	36.9	35.9

Age 50–79

Height in cm (in.)

<160 (<5′3″)	29.3	1.9	33.1	33.6
160–169 (5′3″–5′6″)	31.7	1.6	34.9	35.6
170–179 (5′7″–5′10″)	31.9	1.7	35.3	35.6
≥ 180 (≥5′11″)	33.2	2.5	38.2	36.4
All subjects	30.3	2.4	35.1	35.0

Acceptable Difference

The upper limit of normal increase in latency from one side to the other is 2.0 msec.

Helpful Hints

■ To be certain that the correct nerve fibers are being stimulated, the ankle motion on stimulation should be observed.

- Older persons have a higher incidence of unelicitable H-reflexes than do young persons.
- If the H-reflex is unelicitable on one side, there is a higher likelihood that it will also be unelicitable contralaterally.
- The stimuli are applied every few seconds at random in order to avoid habituation. The subject is asked to relax.
- Jankus and colleagues studied side to side amplitude variability for the tibial H-reflex. They included only subjects with side to side latency differences of less than 1.5 msec. They concluded that a side to side peak to peak amplitude ratio smaller than 0.4 in the face of normal latency is probably abnormal (2).
- The H-reflex can be facilitated with slight active plantarflexion.
- Some of the aforementioned 97th percentile variables followed a slightly aberrant path. For example, in a few cases the older age category actually had a slightly lower 97th percentile value. This was of minimal clinical significance, but the 97th percentile values were adjusted to provide the more conservative cutoff of normal.

Notes

REFERENCES

1. Buschbacher RM: Normal range for H-reflex recording from the calf muscles. *Am J Phys Med Rehabil* 1999; 78:S75–S79.
2. Jankus WR, Robinson LR, Little JW: Normal limits of side to side H-reflex amplitude variability. *Arch Phys Med Rehabil* 1994; 75:3–7.

ADDITIONAL READINGS/ALTERNATE TECHNIQUES

1. Braddom RL, Johnson EW: Standardization of H reflex and diagnostic use in S1 radiculopathy. *Arch Phys Med Rehabil* 1974; 55:161–166.
2. Falco FJE, Hennessey WJ, Goldberg G, Braddom RL: H reflex latency in the healthy elderly. *Muscle Nerve* 1994; 17:161–167.
3. Pease WS, Kozakiewicz R, Johnson EW: Central loop of the H reflex: normal value and use in S1 radiculopathy. *Am J Phys Med Rehabil* 1997; 76:182–184.

CHAPTER 4

Lower Extremity
Sensory and
Mixed Studies

LATERAL FEMORAL CUTANEOUS SENSORY STUDY

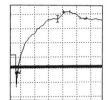

Typical waveform appearance

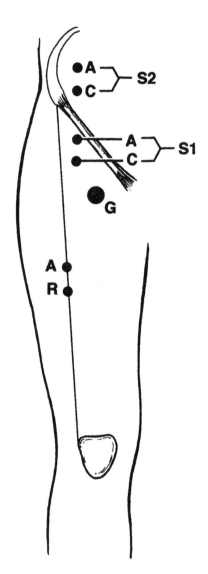

Electrode Placement—Ma and Liveson Technique

Ma and Liveson Technique

Recording electrodes: Surface electrodes are placed along a line connecting the anterior superior iliac spine (ASIS) to the lateral border of the patella with the active electrode (A) 17–20 cm distal to the ASIS and the reference electrode (R) 3 cm more distal.

Ground electrode (G): Placement is between the stimulating and recording electrodes.

Stimulation point 1 (S1): Stimulation can be applied below the inguinal ligament over the origin of the sartorius.

Stimulation point 2 (S2): Stimulation can be applied above the inguinal ligament 1 cm medial to the ASIS.

Machine settings: Standard sensory settings are used.

Nerve fibers tested: L2 and L3 nerve roots through the posterior division of the lumbosacral plexus.

Normal values (1) (20 subjects) (room temperature 23–26 degrees Celsius):

Onset latency (msec)

	Mean	S.D.	Mean + 2 S.D.	Range
S1 (14–18 cm)	2.5	0.2	2.9	2.2–2.8
S2 (17–20 cm)	2.8	0.4	3.6	2.3–3.2

Peak to peak amplitude (µV)

	Mean	S.D.	Range
S1	7.0	1.8	4–11
S2	6.0	1.5	3–10

Helpful Hints

- This study is technically difficult, especially in overweight persons. Absent responses are of questionable clinical significance.
- An alternate technique of recording can be used by drawing digital ring electrodes "flat" and using them as strap or band electrodes.
- For S2 stimulation it may help to exert pressure toward the ASIS. Rotating the anode may be necessary to reduce stimulus artifact.

202   *Lower Extremity Sensory and Mixed Studies*

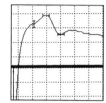

Typical waveform appearance

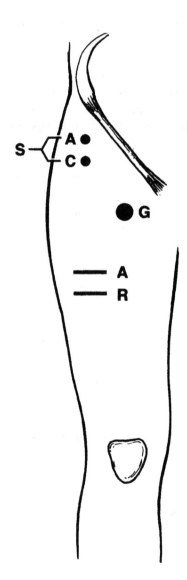

Electrode Placement—Spevak and Prevec Technique

Spevak and Prevec Technique

Recording electrodes (R): Two 8 cm long strip electrodes are placed 2.5 cm apart on the anterolateral thigh approximately 25 cm distal to the stimulating electrode.

Ground electrode (G): Placement is between the stimulating and recording electrodes.

Stimulation point (S): Stimulation is applied 6–10 cm below the ASIS. The point where a sensation radiates to the lateral thigh is sought. Stimulus duration is 0.1 msec with the intensity set to double the sensory threshold and not above 150 V. Eight to 32 responses are averaged.

Machine settings: Sensitivity—1–2 µV/division, Low frequency filter—100 Hz, High frequency filter—5 kHz, Sweep speed—1 msec/division.

Normal values (2) (29 subjects—distance 25.3 ± 3.5 cm) (skin temperature over the thigh 29–32.5 degrees Celsius):

Onset latency (msec)

Mean	S.D.	Mean + 2 S.D.
4.1	0.7	5.5

Peak latency (msec)

Mean	S.D.	Mean + 2 S.D.
4.6	0.7	6.0

Nerve conduction velocity (m/sec) calculated for onset latency

Mean	S.D.	Mean – 2 S.D.
62.3	5.5	51.3

Peak to peak amplitude (µV)

Mean	S.D.
2.0	1.0

Duration (msec)

Mean	S.D.	Mean + 2 S.D.
1.9	0.5	2.9

Acceptable Differences

The side to side difference in conduction velocity is 2.6 ± 2.2 m/sec and never exceeded 6 m/sec in the cited study.

The side to side difference in amplitude is 0.86 ± 0.89 µV.

Helpful Hints

- If the response is unobtainable, it is of doubtful clinical significance. If present, slowing of conduction velocity or a greater than normal side to side conduction velocity difference are most sensitive to pathology.
- This study will be normal in cases of purely local slowing across the inguinal ligament segment.

Notes

REFERENCES

1. Ma DM, Liveson JA: *Nerve conduction handbook.* Philadelphia: FA Davis, 1983.
2. Spevak MK, Prevec TS: A noninvasive method of neurography in meralgia paresthetica. *Muscle Nerve* 1995; 18:601–605.

ADDITIONAL READINGS/ALTERNATE TECHNIQUES

1. Stevens A, Rosselle N: Sensory nerve conduction velocity of n. cutaneus femoris lateralis. *Electromyography* 1970; 4:397–398.
2. Butler ET, Johnson EW, Kaye ZA: Normal conduction velocity in the lateral femoral cutaneous nerve. *Arch Phys Med Rehabil* 1974; 55:31–32.
3. Sarala PK, Nishihara T, Oh SJ: Meralgia paresthetica: electrophysiologic study. *Arch Phys Med Rehabil* 1979; 60:30–31.
4. Lysens R, Vandendriessche G, VanMol Y, Rosselle N: The sensory conduction velocity in the cutaneous femoris lateralis nerve in normal adult subjects and in patients with complaints suggesting meralgia paresthetica. *Electromyogr clin Neurophysiol* 1981; 21:505–510.
5. Karandreas N, Papatheodorou A, Triantaphilos I, et al: Sensory nerve conduction studies of the less frequently examined nerves. *Electromyogr clin Neurophysiol* 1995; 35:169–173.

MEDIAL CALCANEAL SENSORY STUDY

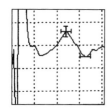

Typical waveform appearance

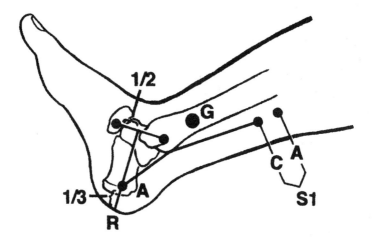

Electrode Placement

Active electrode (A): Placement is one third of the distance from the apex of the heel to a point midway between the navicular tubercle and the prominence of the medial malleolus.

Reference electrode (R): Placement is over the apex of the heel.

Ground electrode (G): Placement is between the stimulating and recording electrodes.

Stimulation point (S): The cathode (C) is placed 10 cm proximal to the active electrode, measuring first to the posterior tip of the medial malleolus and then along the medial border of the tibia. The cathode is placed 1–2 cm posterior to the medial edge of the tibia. The anode (A) is proximal or rotated to minimize stimulus artifact.

Machine settings: Sensitivity—10–20 µV/division, Low frequency filter—2 Hz, High frequency filter—2 kHz, Sweep speed—1 msec/division.

Nerve fibers tested: S1 nerve root through the anterior division of the lumbosacral plexus and the tibial nerve.

Normal values (1) (36 subjects) (skin temperature greater than or equal to 31 degrees Celsius):

Onset latency (msec)

Mean	Mean + 2 S.D.	Range
1.7	2.0	1.4–2.0

Peak latency (msec)

Mean	Mean + 2 S.D.	Range
2.5	2.8	2.2–2.8

Baseline to peak amplitude (µV)

Mean	Range	Lower limit of normal (corrected for skewing)
18	8–34	8

Acceptable Differences

The upper limit of normal side to side difference in onset latency is 0.3 msec.

The upper limit of normal side to side difference in peak latency is 0.3 msec.

The upper limit of normal side to side difference in amplitude is 12 µV.

Helpful Hints

- This sensory response may need to be averaged.
- The sensory response is often followed by volume conducted motor artifact.
- To derive the above normal values, the authors prepared the active and reference electrode sites with abrasive tape and cleansed the skin with alcohol.

Notes

REFERENCE

1. Park TA, DelToro DR: The medial calcaneal nerve: anatomy and nerve conduction technique. *Muscle Nerve* 1995; 18:32–38.

MEDIAL FEMORAL CUTANEOUS SENSORY STUDY

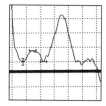

Typical waveform appearance

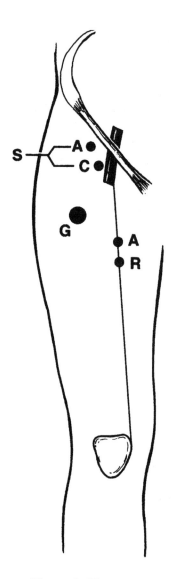

Electrode Placement

Active electrode (A): The active electrode is placed 14 cm distal to the femoral pulse in the inguinal area, along an imaginary line from the pulse to the medial border of the patella.

Reference electrode (R): The reference electrode is placed 4 cm distal to the active electrode on this same imaginary line.

Ground electrode (G): Placement is proximal to the active electrode on the lateral thigh.

Stimulation point (S): The cathode (C) is placed immediately lateral to the femoral artery in the inguinal area. The anode (A) is proximal.

Machine settings: Sensitivity—5 µV/division, Low frequency filter—20 Hz, High frequency filter—2 kHz, Sweep speed—1 msec/division, Stimulator pulse duration—0.2 msec.

Nerve fibers tested: L2 and L3 nerve roots through the posterior division of the lumbosacral plexus and the femoral nerve.

Normal values (1) (32 subjects) (temperature over the mid-medial thigh of 33 degrees Celsius or more):

Onset latency (msec)

Mean	S.D.	Mean + 2 S.D.	Range
2.3-2.4	0.2	2.7-2.8	1.9–2.9

Peak latency (msec)

Mean	S.D.	Mean + 2 S.D.	Range
2.9	0.2-0.3	3.3-3.5	2.3–3.5

Onset to peak amplitude (µV)

Mean	S.D.	Range
4.8-4.9	1.0	3.4–7.9

Helpful Hints

- Averaging of approximately five to 10 recordings is often necessary.
- Anode rotation is often required to reduce stimulus artifact.
- The stimulating electrodes may need to be moved medially or laterally to obtain a recording.
- The leg should be relaxed with the knee slightly flexed. It may be useful to support the knee with a pillow.

Notes

REFERENCE

1. Lee HJ, Bach JR, DeLisa JA: Medial femoral cutaneous nerve conduction. *Am J Phys Med Rehabil* 1995; 74:305–307.

PERONEAL NERVE

DEEP PERONEAL SENSORY STUDY

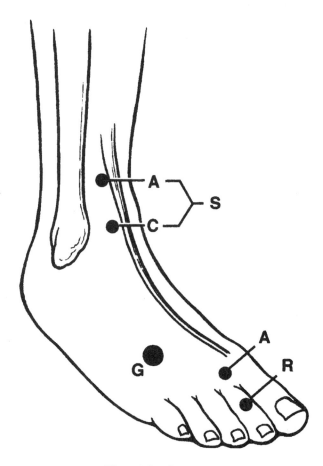

Electrode Placement

Active electrode (A): Placement is over the terminal sensory branch of the nerve at the interspace between the 1st and 2nd metatarsal heads.

Reference electrode (R): Placement is 3 cm distal to the active electrode on the 2nd digit.

Ground (G): Placement is between the stimulating and recording electrodes on the dorsum of the foot.

Stimulation point (S): The cathode (C) is placed at the ankle, 12 cm proximal to the active electrode and just lateral to the extensor hallucis longus tendon. The anode (A) is proximal.

Machine settings: Averager is used (5–20 stimuli). Sensitivity—5 µV/division, Low frequency filter—20 Hz, High frequency filter—2 kHz, Sweep speed—1 msec/division.

Nerve fibers tested: L5 nerve root, through the posterior division of the lumbosacral plexus and the common and deep peroneal nerves.

Normal values (1) (40 subjects) (skin temperature over the dorsum of the foot greater than or equal to 29 degrees Celsius):

Onset latency (msec)

Mean	S.D.	Mean + 2 S.D.	Range
2.9	0.4	3.7	2.1–3.6

Peak latency (msec)

Mean	S.D.	Mean + 2 S.D.	Range
3.6	0.4	4.4	2.7–4.2

Onset to peak amplitude (µV)

Mean	S.D.	Range
3.4	1.2	1.6–6.6

Notes

REFERENCE

1. Lee HJ, Bach JR, DeLisa JA: Deep peroneal sensory nerve: standardization in nerve conduction study. *Am J Phys Med Rehabil* 1990; 69:202–204.

ADDITIONAL READINGS/ALTERNATE TECHNIQUES

1. Posas HN Jr, Rivner MH: Nerve conduction studies of the medial branch of the deep peroneal nerve. *Muscle Nerve* 1990; 13:862.
2. Ponsford S: Medial (cutaneous) branch of deep common peroneal nerve: recording technique and a case report. *Electroencephalogr clin Neurophysiol* 1994; 93:159–160.
3. Weimer LH, Trojaborg W, Marquinez AI, et al: The deep peroneal sensory nerve: a reliable source of nerve conduction data? Presented at the AAEM annual meeting, Orlando, Florida, October 17, 1998.

SUPERFICIAL PERONEAL SENSORY STUDY (MEDIAL AND INTERMEDIATE DORSAL CUTANEOUS BRANCHES)

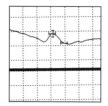

Typical waveform appearance

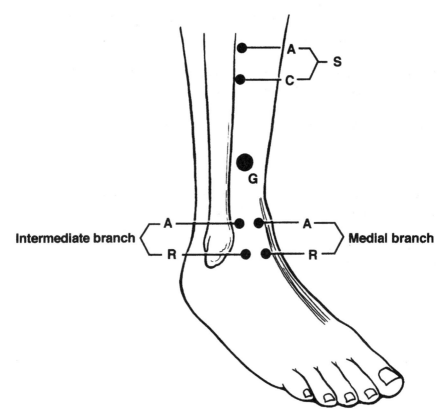

Electrode Placement

Izzo et al. Technique

Active electrodes (A): Placement is at the level of the ankle after localization by inspection and palpation during plantar flexion and inversion. The medial branch passes over the anterior ankle to the dorsum of the foot. It lies just lateral to the tendon of the extensor hallucis longus. The intermediate branch lies 1–2 cm medial to the lateral malleolus.

Reference electrode (R): Placement is 3–4 cm distal to the active electrode for both branches (1,2).

Ground electrode (G): Placement is over the distal dorsal lower leg, between the active electrodes and the cathode.

Stimulation point (S): The cathode (C) is placed 14 cm proximal to the active electrode on the anterolateral aspect of the lower leg. The anode (A) is proximal.

Machine settings: Sensitivity—20 µV/division, Low frequency filter—20 Hz, High frequency filter—2 kHz, Sweep speed—2 msec/division.

Nerve fibers tested: L4, L5, and S1 nerve roots, through the posterior division of the lumbosacral plexus and the common peroneal nerve.

Normal values (1) (80 subjects) (ankle skin temperature of at least 28 degrees Celsius):

Onset latency (msec)

Medial Branch

Mean	S.D.	Mean + 2 S.D.	Range
2.8	0.3	3.4	2.2–3.6

Intermediate Branch

Mean	S.D.	Mean + 2 S.D.	Range
2.8	0.3	3.4	2.2–3.6

Peak latency (msec)

Medial Branch

Mean	S.D.	Mean + 2 S.D.	Range
3.4	0.4	4.2	2.7–4.7

Intermediate Branch

Mean	S.D.	Mean + 2 S.D.	Range
3.4	0.4	4.2	2.8–4.6

Peak to peak amplitude (μV)

Medial Branch

Mean	S.D.	Range
18.3	8.0	5–44

Intermediate Branch

Mean	S.D.	Range
15.1	8.2	4–40

Data for 122 elderly subjects (unobtainable in 12), aged 60–89 years, mean 74.1, with skin temperature over the extensor digitorum brevis at least 29 degrees Celsius (2).

Onset latency (msec)

Medial Branch

Mean	S.D.	Mean + 2 S.D.
3.1	0.3	3.7

Onset to peak amplitude (µV)

Medial Branch

Mean	S.D.
7.7	3.9

Helpful Hints

- These nerve branches are superficial and can often be palpated or seen passing under the surface of the skin, especially when the foot is plantarflexed and inverted (1).
- A stimulus duration of 0.05–0.1 msec was used to derive the data presented by Izzo and colleagues (1).
- In approximately 2 percent of subjects a response is unobtainable from the intermediate branch (1).

Jabre Technique (Intermediate Branch)

Recording electrodes: A bar electrode is placed at the level of the lateral malleolus, one to two fingerbreadths medial to the malleolus. The active electrode (A) is proximal and the reference electrode (R) is distal (3,4).

Ground electrode (G): The ground electrode is placed over the anterior lower leg, between the stimulating and recording electrodes.

Stimulation point (S): The cathode (C) is placed 12 cm proximal to the active electrode with the stimulator probe held firmly against the anterior aspect of the fibula. The anode (A) is proximal.

Machine settings: Sensitivity—10 μV/division, Low frequency filter—32 Hz, High frequency filter—1.6 kHz, Sweep speed—2 msec/division.

Normal values (3) (36 subjects) (data derived at an ambient temperature of 70 degrees Fahrenheit):

Peak latency (msec)

Mean	S.D.	Mean + 2 S.D.
2.9	0.3	3.5

Onset to negative peak amplitude (μV)

Mean	S.D.
20.5	6.1

Helpful Hints

- In some cases the nerve can be palpated for easier localization.
- A relatively low stimulus intensity may be necessary to avoid contamination by motor artifact.
- The data reported here were derived using a 0.05 msec pulse duration. A 0.1 msec pulse duration may be required in some cases.
- A second, more proximal stimulation allows for calculation of a conduction velocity. Stimulation is applied 8–9 cm proximal to the above-described stimulation point. The conduction velocity of 17 subjects was 65.7 ± 3.7 m/sec between these points (3).

Notes _____

_____

_____

_____

_____

_____

_____

REFERENCES

1. Izzo KL, Sridhara CR, Rosenholtz H, Lemont H: Sensory conduction studies of the branches of the superficial peroneal nerve. *Arch Phys Med Rehabil* 1981; 62:24–27.
2. Falco FJE, Hennessey WJ, Goldberg G, Braddom RL: Standardized nerve conduction studies in the lower limb of the healthy elderly. *Am J Phys Med Rehabil* 1994; 73:168–174.
3. Jabre JF: The superficial peroneal sensory nerve revisited. *Arch Neurol* 1981; 38:666–667.
4. Jabre JF, Hackett ER: *EMG manual.* Springfield, Ill.: Charles C. Thomas, 1983.

ADDITIONAL READINGS/ALTERNATE TECHNIQUES

1. DiBenedetto M: Sensory nerve conduction in lower extremities. *Arch Phys Med Rehabil* 1970:253-258.
2. Behse F, Buchthal F: Normal sensory conduction in the nerves of the leg in man. *J Neurol Neurosurg Psychiatry* 1971; 34:404–414.
3. Cape CA: Sensory nerve action potentials of the peroneal, sural and tibial nerves. *Am J Phys Med* 1971; 50:220–229.
4. Ma DM, Liveson JA: *Nerve conduction handbook.* Philadelphia: FA Davis, 1983.
5. Karandreas N, Papatheodorou A, Triantaphilos I, et al: Sensory nerve conduction studies of the less frequently examined nerves. *Electromyogr clin Neurophysiol* 1995; 35:169–173.

POSTERIOR FEMORAL CUTANEOUS SENSORY STUDY

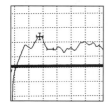

Typical waveform appearance

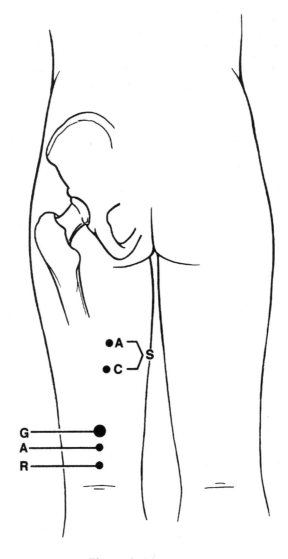

Electrode Placement

Recording electrodes: A bar electrode is placed at the midline of the posterior thigh, with the active electrode (A) 6 cm proximal to the midpopliteal region. The reference (R) electrode is distal.

Ground electrode (G): Placement is just proximal to the bar electrode.

Stimulation point (S): The cathode (C) is placed 12 cm proximal to the active electrode on a line connecting the active electrode with the ischial tuberosity, in the groove between the medial and lateral hamstring musculature (the intermuscular groove can be palpated by having the subject flex the knee. The anode (A) is proximal.

Machine settings: Sensitivity—5 µV/division, Low frequency filter—20 Hz, High frequency filter—2 kHz, Sweep speed—1–2 msec/division.

Nerve fibers tested: Posterior divisions of the S1 and S2 nerve roots and anterior divisions of the S2 and S3 nerve roots.

Normal values (1) (40 subjects) (skin temperature of the posterior thigh maintained between 32 and 33 degrees Celsius):

Peak latency (msec)

Mean	S.D.	Mean + 2 S.D.	Range
2.8	0.2	3.2	2.3–3.3

Peak to peak amplitude (µV)

Mean	S.D.	Range
6.5	1.5	4.1–12.0

Helpful Hint

- Local depolarization of the surrounding musculature may occur on stimulation. This does not generally obscure the sensory waveform but is a theoretical confounding factor.

Notes

REFERENCE

1. Dumitru D, Nelson MR: Posterior femoral cutaneous nerve conduction. *Arch Phys Med Rehabil* 1990; 71:979–982.

SAPHENOUS NERVE

SAPHENOUS SENSORY STUDY (DISTAL TECHNIQUE)

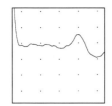

Typical waveform appearance

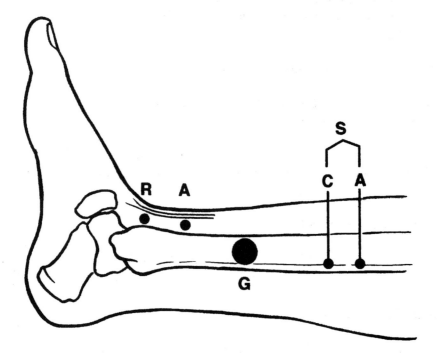

Electrode Placement

Recording electrodes: A 3 cm bar electrode is used (1). The reference electrode (R) is positioned slightly anterior to the highest prominence of the medial malleolus, between the malleolus and the tendon of the tibialis anterior. The active electrode (A) is proximal and slightly medial to the tibialis anterior tendon.

Ground electrode (G): Placement is between the stimulating and recording electrodes.

Stimulation point (S): The cathode (C) is placed 14 cm proximal to the active electrode deep to the medial border of the tibia. The anode (A) is proximal.

Machine settings: Sensory settings are used.

Nerve fibers tested: L3 and L4 nerve roots, through the posterior division of the lumbosacral plexus. This nerve is a continuation of the femoral nerve.

Normal values (2) (80 subjects) (skin temperature at the ankle greater than or equal to 28 degrees Celsius):

Onset latency (msec)

Mean	S.D.	Mean + 2 S.D.	Range
2.9	0.4	3.7	2.1–3.8

Peak latency (msec)

Mean	S.D.	Mean + 2 S.D.	Range
3.5	0.4	4.3	2.7–4.8

Peak to peak amplitude (µV)

Mean	S.D.	Range
5.4	2.5	1–15

Helpful Hints

- Small amplitudes are common.
- Firm pressure should be applied to the stimulator. The plantar flexors should be relaxed and the ankle can be placed in slight plantarflexion (1).
- Averaging may be necessary.
- This response is often unrecordable.
- A sweep speed of 5 msec/division has been described to improve visualization of the waveform with low amplitude potentials. (1)

Notes

REFERENCES

1. Wainapel SF, Kim DJ, Ebel A: Conduction studies of the saphenous nerve in healthy subjects. *Arch Phys Med Rehabil* 1978; 59:316–319.
2. Izzo KL, Sridhara CR, Rosenholtz H, Lemont H: Sensory conduction studies of the branches of the superficial peroneal nerve. *Arch Phys Med Rehabil* 1981; 62:24–27.

SAPHENOUS SENSORY STUDY (PROXIMAL TECHNIQUE)

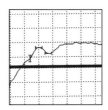

Typical waveform appearance

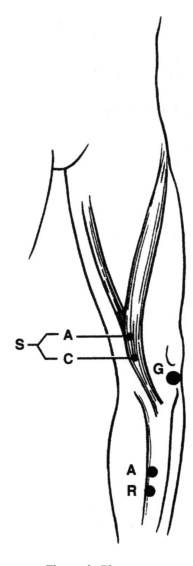

Electrode Placement

Recording electrodes: The active electrode (A) is placed 15 cm distal to the cathode on the medial border of the tibia. The reference electrode (R) is placed 3 cm distally. A 3 cm bar electrode can be used.

Ground electrode (G): Placement is between the stimulating and recording electrodes.

Stimulation point (S): The knee is slightly flexed. The cathode (C) is placed on the medial knee between the tendons of the sartorius and gracilis muscles, approximately 1 cm above the level of the inferior border of the patella. The anode (A) is proximal.

Machine settings: Sensory settings are used.

Nerve fibers tested: L3 and L4 nerve roots, through the posterior division of the lumbosacral plexus. This nerve is a continuation of the femoral nerve.

Normal values (1) (28 subjects—over a 13–16 cm distance) (room temperature 23–26 degrees Celsius):

Onset latency (msec)

Mean	S.D.	Mean + 2 S.D.	Range
2.5	0.19	2.88	2.2–2.8

Peak to peak amplitude (µV)

Mean	S.D.	Range
10.23	2.05	7.0–15.0

Helpful Hints

- This nerve may be difficult to localize, especially in obese persons. The hamstring tendon can be palpated at the posterior aspect of the medial knee. Anterior to this is the gracilis, then the sartorius.

- If stimulation is performed too far anteriorly, subjects may report a sensation in the patellar region. They should feel paresthesias in the medial foreleg to the ankle.
- Firm pressure should be applied to the stimulating and recording electrodes.
- A study of the infrapatellar branch of the saphenous nerve has also been described (2). Twenty-five subjects were tested using an active electrode at the inferior medial edge of the patella with the reference electrode on the lower edge of the patella. Stimulation was applied just superior to the medial epicondyle of the femur between the sartorius and gracilis muscles. Onset latency was 1.56 ± 0.3 msec. Peak latency was 1.9 ± 0.2 msec. Baseline to peak amplitude was 4 ± 2.1 µV. Maximum side to side difference was 0.4 msec for latency and 3.7 µV for amplitude. A response was obtained in 90% of subjects after averaging 20 recordings.

Notes

REFERENCES

1. Ma DM, Liveson JA: *Nerve conduction handbook.* Philadelphia: FA Davis, 1983.
2. Gutierrez JE, Ordonez V: Normal sensory conduction in the infrapatellar branch of the saphenous nerve. Presented at the AAEM annual meeting, September 20, 1997.

SURAL NERVE

SURAL LATERAL DORSAL CUTANEOUS BRANCH SENSORY STUDY

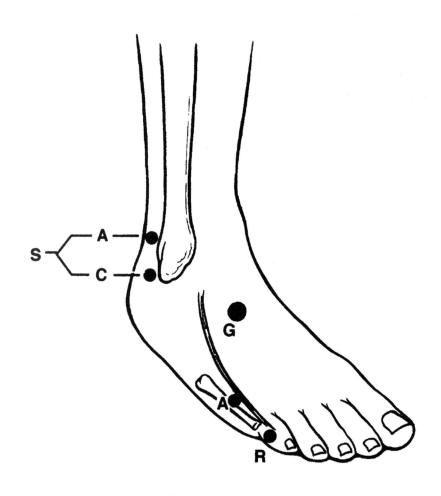

Electrode Placement

Recording electrodes: Felt-tip electrodes with a fixed interelectrode distance of 37 mm were used in the cited study. Placement is such that the active electrode (A) is over the dorsolateral surface of the foot at the midpoint of the 5th metatarsal and just lateral to the extensor digitorum brevis tendon of the 5th toe. The reference electrode (R) is distal.

Ground (G): Placement is on the dorsum of the foot.

Stimulation point (S): The cathode (C) is placed 12 cm proximal to the active electrode behind the lateral malleolus. The anode (A) is proximal.

Machine settings: Averager is used (5–10 stimuli). Sensitivity—5 µV/division, Low frequency filter—20 Hz, High frequency filter—2 kHz, Sweep speed—1 msec/division.

Nerve fibers tested: S1 and S2 nerve roots, through the anterior and posterior divisions of the lumbosacral plexus and the tibial and peroneal nerves.

Normal values (1) (40 subjects) (skin temperature over the dorsum of the foot greater than or equal to 31 degrees Celsius):

Onset latency (msec)

Mean	S.D.	Mean + 2 S.D.	Range
3.2	0.4	4.0	2.5–4.0

Peak latency (msec)

Mean	S.D.	Mean + 2 S.D.	Range
3.9	0.5	4.9	3.0–4.9

Baseline to peak amplitude (µV)

Mean	S.D.	Range
5.8	2.1	3.0–11.0

Notes

REFERENCE

1. Lee HJ, Bach HJR, DeLisa JA: Lateral dorsal cutaneous branch of the sural nerve: standardization in nerve conduction study. *Am J Phys Med Rehabil* 1992; 71:318–320.

SURAL SENSORY STUDY

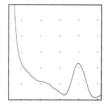

Typical waveform appearance

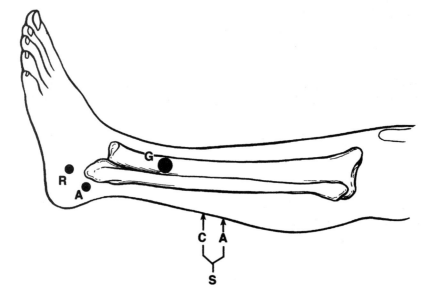

Electrode Placement

Recording electrodes: Placement of the active electrode (A) is behind the lateral malleolus with the reference electrode (R) 3 cm distal. A 3 cm bar electrode is commonly used.

Ground electrode (G): Placement is between the stimulating and recording electrodes.

Stimulation point (S): The cathode (C) is placed 14 cm proximal to the active electrode in the midline or slightly lateral to the midline of the posterior lower leg. The anode (A) is proximal.

Machine settings: Sensitivity—20 µV/division, Low frequency filter—20 Hz, High frequency filter—2 kHz, Sweep speed—2 msec/division (1,2).

Nerve fibers tested: S1 and S2 nerve roots, through the anterior and posterior divisions of the lumbosacral plexus and the tibial and peroneal nerves.

Normal values (1) (80 subjects—absent in 2) (skin temperature over the ankle at least 28 degrees Celsius):

Onset latency (msec)

Mean	S.D.	Mean + 2 S.D.	Range
2.9	0.3	3.5	2.3–3.7

Peak latency (msec)

Mean	S.D.	Mean + 2 S.D.	Range
3.6	0.4	4.4	2.8–4.4

Peak to peak amplitude (µV)

Mean	S.D.	Range
16.6	7.5	5–56

Data for 122 elderly subjects (unobtainable in 4), aged 60–89 years, mean 74.1, with skin temperature over the extensor digitorum brevis at least 29 degrees (2)

Onset latency (msec)

Mean	S.D.	Mean + 2 S.D.
3.2	0.4	4.0

Onset to peak amplitude (μV)

Mean	S.D.
10.3	5.8

Helpful Hints

- The stimulator may need to be moved slightly to one side or the other to obtain a response.
- The technique is best performed with the subject lying on his side.
- Occasionally the responses may need to be averaged.

Notes _____

REFERENCES

1. Izzo KL, Sridhara CR, Rosenholtz H, Lemont H: Sensory conduction studies of the branches of the superficial peroneal nerve. *Arch Phys Med Rehabil* 1981; 62:24–27.
2. Falco FJE, Hennessey WJ, Goldberg G, Braddom RL: Standardized nerve conduction studies in the lower limb of the healthy elderly. *Am J Phys Med Rehabil* 1994; 73:168–174.

ADDITIONAL READINGS/ALTERNATE TECHNIQUES

1. DiBenedetto M: Sensory nerve conduction in lower extremities. *Arch Phys Med Rehabil* 1970; 253–258.
2. Cape CA: Sensory nerve action potentials of the peroneal, sural and tibial nerves. *Am J Phys Med* 1971; 50:220–229.
3. Schuchmann JA: Sural nerve conduction: a standardized technique. *Arch Phys Med Rehabil* 1977; 58:166–168.
4. Wainapel SF, Kim DJ, Ebel A: Conduction studies of the saphenous nerve in healthy subjects. *Arch Phys Med Rehabil* 1978; 59:316–319.
5. Truong XT, Russo FI, Vagi I, Rippel DV: Conduction velocity in the proximal sural nerve. *Arch Phys Med Rehabil* 1979; 60:304–308.
6. Horowitz SH, Krarup C: Conduction studies of the normal sural nerve. *Muscle Nerve* 1992; 15:374–383.

TIBIAL NERVE COMPOUND ACTION POTENTIALS (MEDIAL AND LATERAL PLANTAR NERVES)

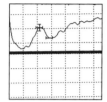

Typical waveform appearance

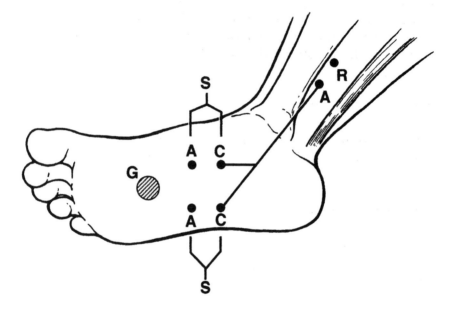

Electrode Placement

Recording electrodes: A bar electrode is placed over the tibial nerve just proximal to the flexor retinaculum (proximal to a line from the posterior calcaneous to the medial malleolus. The active electrode (A) is distal and the reference electrode (R) is proximal. (1)

Ground electrode (G): Placement is over the dorsum of the foot.

Stimulation points (S): The cathode (C) is placed 14 cm distal to the active recording electrode as shown in the accompanying figure for both the medial and lateral plantar branches. The anode (A) is distal. For the medial branch the distance is measured 10 cm to the interspace between the 1st and 2nd metatarsals and then 4 cm distally. For the lateral branch the stimulation site is between the 4th and 5th metatarsals (1)

Machine settings: Routine sensory settings are used.

Normal values (2) (41 subjects) (skin temperature over the tarsal tunnel and the medial and lateral sole of the foot 26–32 degrees Celsius):

Peak latency (msec)

	Mean	S.D.	Mean + 2 S.D.	Range
Medial	3.16	0.26	3.68	2.6–3.7
Lateral	3.15	0.25	3.65	2.7–3.7

Amplitude (µV)

	Range
Medial	10–30
Lateral	8–20

Helpful Hints

- Although technically a mixed nerve study, this technique approximates a sensory study.
- Firm pressure should be exerted on the stimulating and recording electrodes.
- Stimulus artifact may interfere with the recording, especially in persons with thick plantar skin.
- This response is often difficult to elicit even in normal subjects. Unelicitable waveforms must be interpreted with caution.

Notes

REFERENCES

1. Dumitru D: Nerve conduction studies. In: Dumitru D, *Electrodiagnostic medicine*. Philadelphia: Hanley and Belfus, 1995.
2. Saeed MA, Gatens PF: Compound nerve action potentials of the medial and lateral plantar nerves through the tarsal tunnel. *Arch Phys Med Rehabil* 1982; 63:304–307.

ADDITIONAL READINGS/ALTERNATE TECHNIQUES

1. Oh SJ, Sarala PK, Kuba T, Elmore RS: Tarsal tunnel syndrome: electrophysiological study. *Ann Neurol* 1979; 5:327–330.
2. Cape CA: Sensory nerve action potentials of the peroneal, sural and tibial nerves. *Am J Phys Med* 1971; 50:220–229.

Chapter 5

Cranial Nerves

BLINK REFLEX

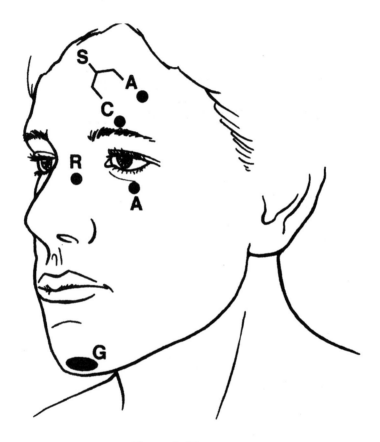

Electrode Placement

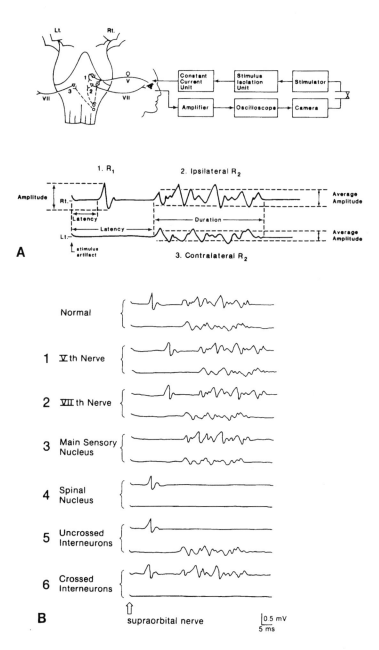

From ELECTRODIAGNOSIS IN DISEASES OF NERVE AND MUSCLE by Jun Kimura. Copyright © 1989 by Oxford University Press, Inc. Used by permission of Oxford University Press, Inc.

Active electrode (A): Placement is over the lower lateral orbicularis oculi muscle bilaterally (1).

Reference electrode (R): Placement is over the temple or the lateral surface of the nose above the nasalis muscle (1,2).

Ground electrode (G): Placement is under the chin or on the forehead or cheek (1,2).

Stimulation point (S): Stimulation is applied with the cathode (C) over the supraorbital nerve at the supraorbital notch. The anode (A) is superiolateral. (Direct stimulation can also be applied to the facial nerve—see section on CN VII.) (1,2).

Machine settings: Sensitivity—50–200 µV, Low frequency filter—20 Hz, High frequency filter—10 kHz, Sweep speed—10 msc/division (1,2).

Nerve fibers tested: Afferent CN V and efferent CN VII fibers, as well as their central connections.

Normal values (1) (83 subjects):

Direct latency (D) to facial nerve stimulation (msec)

Mean	S.D.	Upper Limit of Normal (Mean + 3 S.D.)
2.9	0.4	4.1

R1 latency (msec)

Mean	S.D.	Upper Limit of Normal (Mean + 3 S.D.)
10.5	0.8	13.0

Ipsilateral R2 latency (msec)

Mean	S.D.	Upper Limit of Normal (Mean + 3 S.D.)
30.5	3.4	40

Contralateral R2 latency (msec)

Mean	S.D.	Upper Limit of Normal (Mean + 3 S.D.)
30.5	4.4	41

Acceptable Differences

The upper limit of normal side to side difference in direct response latency (D) is 0.6 msec.

The upper limit of normal side to side difference in R1 latency is 1.2 msec (3 S.D.).

The upper limit of normal side to side difference in R2 latency evoked by stimulation on one side (dual channel recording) is 5 msec. If the R2 is recorded by first stimulating one side and then the other (single channel recording), the upper limit of normal difference is 7 msec.

Helpful Hints

- The subject should be relaxed.
- The R2 latency may be variable.
- Several blink responses should be tested and the shortest latencies chosen. Kimura recommends that at least eight trials be performed (1).
- Latency should be measured to the first deflection from the baseline.
- Excessive numbers of stimuli may lead to habituation and should be avoided.
- A ratio of R1 latency (R) to the direct response latency (D) to facial nerve stimulation (see section on CN VII) can be calculated. This R/D ratio should not fall outside the range of 2.6 to 4.6 (1). A larger ratio, with a normal D, is indicative of slowing of the trigeminal portion of the arc. If the R/D ratio is low, it is indicative of slowing of the facial nerve component.

- Care should be taken to aim the anode away from the contralateral side so that bilateral stimulation is not performed inadvertently.
- In 5–10% of normal subjects paired stimuli are necessary to record a stable R1 (1).
- A glabellar tap with a special hammer can be used instead of electrical stimulation.

Notes

REFERENCES

1. Kimura J: *Electrodiagnosis in diseases of nerve and muscle: principles and practice,* 2nd ed. Philadelphia: FA Davis, 1989.
2. Dumitru D: *Electrodiagnostic medicine.* Philadelphia: Hanley and Belfus, 1995.

ADDITIONAL READINGS/ALTERNATE TECHNIQUES

1. Kimura J, Powers JM, Van Allen MW: Reflex response of orbicularis oculi muscle to supraorbital nerve stimulation. *Arch Neurol* 1969; 21:193–199.
2. Kimura J, Bodensteiner J, Yamada T: Electrically elicited blink reflex in normal neonates. *Arch Neurol* 1977; 34:246–249.
3. Ma DM, Liveson JA: *Nerve conduction handbook.* Philadelphia: FA Davis, 1983.
4. Dumitru D, Walsh NE, Porter LD: Electrophysiological evaluation of the facial nerve in Bell's palsy. *Am J Phys Med Rehabil* 1988; 67:137–144.
5. Ellrich J, Hopf HC: The R3 component of the blink reflex: normative data and application in spinal lesions. *Electroencephalogr clin Neurophysiol* 1996; 101:349–354.

CRANIAL NERVE VII

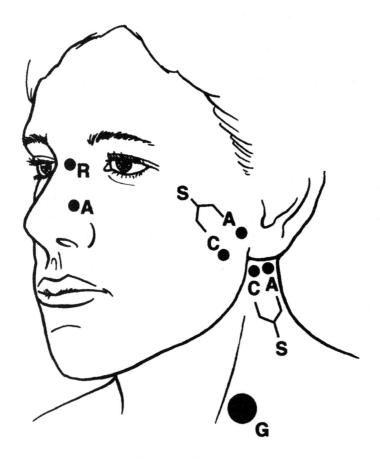

Electrode Placement

Active electrode (A): Placement can be over the nasalis, orbicularis oris, orbicularis oculi, and levator labii superioris (or essentially any muscle innervated by the nerve). For recording from the nasalis, the active electrode is placed on the lateral mid-nose. The subject may "wrinkle the nose" and the electrode is placed on the most prominent bulge of the muscle. For the orbicularis oculi the active electrode is placed under the eye in line with the pupil. An alternate position is at the lateral border of the eye. For the orbicularis oris the active electrode is placed lateral to the angle of the mouth (1-4).

Reference electrode (R): Placement is on the tip or bridge of the nose (1,2).

Ground electrode (G): Placement is over the base of the neck or on the cheek (1,2).

Stimulation point (S): Preauricular stimulation is performed with the cathode (C) just anterior to the lower ear over the substance of the parotid gland and several centimeters superior to the angle of the mandible. Postauricular stimulation is performed by placing the cathode just behind the lower ear, below the mastoid process and behind the neck of the mandible. The anode (A) is posterior (1,2).

Machine settings: Sensitivity—200–1000 µV, Low frequency filter—8 Hz, High frequency filter—8 kHz, Sweep speed—1–2 msec/division (2).

Nerve fibers tested: CN VII efferent motor fibers.

Normal values (1) (44 subjects) (room temperature 23–26 degrees Celsius):

Preauricular stimulation latency to the nasalis (msec)

Mean	S.D	Range
3.57	0.35	2.8–4.1

Postauricular stimulation latency to the nasalis (msec)

Mean	S.D.	Range
3.88	0.36	3.2–4.4

Acceptable Differences

A side to side difference in amplitude of greater than 50% is suggestive of pathology, but the waveform must be similar on both sides (2).

Helpful Hints

- Serial study of the amplitude may be clinically useful.
- Recordings may be made from any CN VII muscle. Surface recording is preferred as this allows assessment of amplitude. The amplitude should be measured from onset to the peak of the negative wave. Needle recording technique has also been described (3,4). With concentric needle recording from the orbicularis oris and stimulation with the cathode over the stylomastoid foramen, the latency in 40 adult subjects was reported as 4.0 ± 0.5 msec (3).
- The motor points in the facial muscles may be poorly defined and an initially negative deflection may not be obtained. If this is the case, latency should be recorded at the initial deflection from baseline (2).
- In cases of CN VII pathology, direct activation of the masseter, especially when recording from orbicularis oris, can give a false volume conducted response. This muscle can be palpated to ensure that this is not occurring. If this is a confounding factor, the facial nerve can be stimulated under the Zygoma. This results in a shorter latency recording, but side to side comparisons can still be made (2).

Notes

REFERENCES

1. Ma DM, Liveson JA: *Nerve conduction handbook.* Philadelphia: FA Davis, 1983.
2. Dumitru D: *Electrodiagnostic medicine.* Philadelphia: Hanley and Belfus, 1995.
3. Taylor N, Jebsen RH, Tenckhoff HA: Facial nerve conduction latency in chronic renal insufficiency. *Arch Phys Med Rehabil* 1970; 51:259–263.
4. Waylonis GW, Johnson EW: Facial nerve conduction delay. *Arch Phys Med Rehabil* 1964; 45:539–547.

ADDITIONAL READINGS/ALTERNATE TECHNIQUES

1. Dumitru D, Walsh NE, Porter LD: Electrophysiological evaluation of the facial nerve in Bell's palsy. *Am J Phys Med Rehabil* 1988; 67:137–144.
2. DeMeirsman J, Claes G, Geerdens L: Normal latency value of the facial nerve with detection in the posterior auricular muscle and normal amplitude value of the evoked action potential. *Electromyogr clin Neurophysiol* 1980; 20:481–485.

CRANIAL NERVE XI

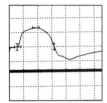

Typical waveform appearance

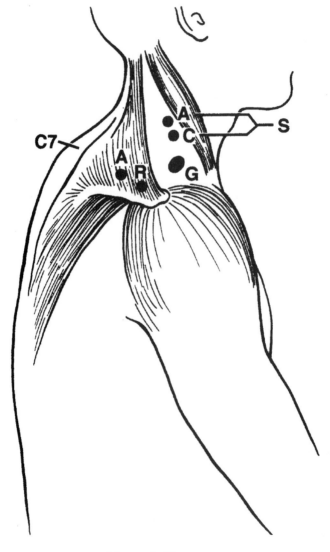

Electrode Placement

Active electrode (A): Placement is over the upper trapezius, about 9 cm lateral to the 7th spinous process.

Reference electrode (R): Placement is 3 cm lateral to the active electrode.

Ground electrode (G): Placement is between the stimulating and recording electrodes.

Stimulation point (S): The cathode (C) is placed in the posterior triangle of the neck, 1–2 cm posterior to the posterior border sternocleidomastoid muscle and slightly above the midpoint of this muscle. This is a point halfway between the mastoid process and the suprasternal notch. The anode (A) is superior (1,2).

Nerve fibers tested: CN XI.

Machine settings: Standard motor settings are used, with a sweep speed of 1–2 msec/division (2).

Normal values (1) (28 subjects) (room temperature 23–26 degrees Celsius):

Onset latency (msec)

Mean	S.D.	range
2.3	0.4	1.7–3.0

Peak to peak amplitude (mV)

>3–4

Helpful Hints

- Care should be taken not to stimulate the brachial plexus.
- The shoulder should shrug with activation of CN XI.

Notes

REFERENCES

1. Ma DM, Liveson JA: *Nerve conduction handbook*. Philadelphia: FA Davis, 1983.
2. Dumitru D: *Electrodiagnostic medicine*. Philadelphia: Hanley and Belfus, 1995.

ADDITIONAL READINGS/ALTERNATE TECHNIQUES

1. Cherington M: Accessory nerve. *Arch Neurol* 1968; 18:708–709.
2. LoMonaco M, DiPasqua PG, Tonali P: Conduction studies along the accessory, long thoracic, dorsal scapular, and thoracodorsal nerves. *Acta Neurol Scand* 1983; 68:171–176.
3. Green RF, Brien M: Accessory nerve latency to the middle and lower trapezius. *Arch Phys Med Rehabil* 1985; 66:23–24.
4. Shankar K, Means KM: Accessory nerve conduction in neck dissection subjects. *Arch Phys Med Rehabil* 1990; 71:403–405.

CHAPTER 6

Root Stimulation

CERVICAL NERVE ROOT STIMULATION

Recording electrodes: Depending on which nerve roots are being tested, placement is over the motor points of the abductor digiti minimi (C8–T1 roots), biceps (C5–C6 roots), or triceps (C6–C8 roots) muscles. The reference electrode is placed over a distal tendon (1).

Stimulation point (S): Stimulation is performed with a 50 mm monopolar needle electrode. This cathode is inserted perpendicular to the skin immediately lateral to the spinous processes so that the tip rests on the vertebral laminae. For the C5/C6 root the needle is placed at the level of the C5 vertebra. For the C6/C7/C8 roots the needle is placed at the level of the C6 vertebra. The C8/T1 roots are stimulated at the level of the C7 vertebra. The anode is a surface electrode and is placed 1 cm caudal and slightly medial to the cathode.

Machine settings: Low frequency filter—8 Hz, High frequency filter—8 kHz.

Acceptable Differences (1) (30 root pairs in 12 subjects)

 The upper limit of normal latency asymmetry from side to side is 1.0 msec.
 The upper limit of normal amplitude asymmetry (percentage reduction from the larger to the smaller value) from side to side is 20%.

Helpful Hints

- Needle recording has also been recommended when recording from the triceps (2). Reliable amplitude measurements can, however, only be made using surface electrode recording.
- A slightly different stimulation procedure has also been studied. Stimulation is similar to that described above, but the monopolar needle electrode is inserted 1–2 cm lateral and inferior to the corresponding spinous processes. The needle is inserted until bone is encountered and is then withdrawn several millimeters. A stimulus duration of 0.05 msec is usually adequate. The anode can be either a surface electrode or another monopolar needle electrode inserted at the same site contralaterally (2-4).

- This is a nonspecific test and does not alone make the diagnosis of radiculopathy. Any other pathology along the route of the nerve can cause slowing, so other pathology needs to be ruled out.
- Side to side comparison is limited if pathology is bilateral.
- A reduction in amplitude of 50% or more between limb stimulation and root stimulation has been used to define a proximal conduction block. The limb stimulation is applied above the common sites of nerve compression (for example, for the ulnar nerve it is applied above the elbow) (5).

Notes

REFERENCES

1. Berger AR, Busis NA, Logigian EL, et al: Cervical root stimulation in the diagnosis of radiculopathy. *Neurology* 1987; 37:329–332.
2. Dumitru D: *Electrodiagnostic medicine.* Philadelphia: Hanley and Belfus, 1995.
3. Kraft GH, Johnson EW: Proximal motor nerve conduction and late responses: an AAEM workshop. American Association of Electrodiagnostic Medicine, 1986.
4. MacLean IC: Spinal nerve stimulation. In AAEM Course B: nerve conduction studies—a review. American Association of Electrodiagnostic Medicine, Rochester, Minnesota, 1988.
5. Menkes DL, Hood DC, Ballesteros RA, Williams DA: Root stimulation improves the detection of acquired demyelinating polyneuropathies. *Muscle Nerve* 1988; 21:298–308.

ADDITIONAL READINGS/ALTERNATE TECHNIQUES

1. Livingstone EF, DeLisa JA, Halar EM: Electrodiagnostic values through the thoracic outlet using C8 root needle studies, F-waves, and cervical somatosensory evoked potentials. *Arch Phys Med Rehabil* 1984; 65:726–730.
2. Evans BA, Daube JR, Litchy WJ: A comparison of magnetic and electrical stimulation of spinal nerves. *Muscle Nerve* 1990; 13:414–420.
3. Tsai CP, Huang CI, Wang V, et al: Evaluation of cervical radiculopathy by cervical root stimulation. *Electromyogr clin Neurophysiol* 1994; 34:363–366.

LUMBAR ROOT STIMULATION

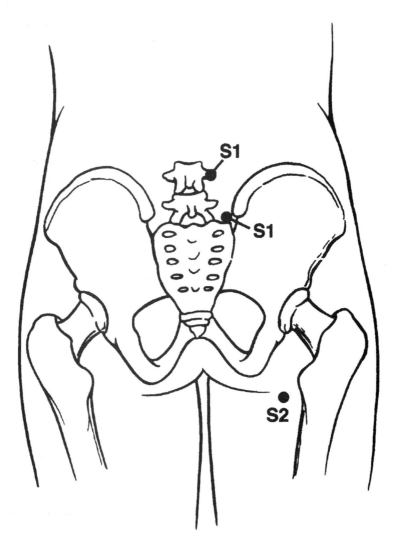

Electrode Placement

Root Stimulation

Recording electrodes: The active and reference electrodes can be placed on any appropriate muscle of the lower extremity (1–3). The active electrode is placed over the motor point or central portion of the muscle. The reference electrode is placed over the distal tendinous insertion of that muscle. Needle recording may be used.

Ground electrode: Placement is between the stimulating and recording electrodes.

Stimulation point 1 (S1): A 50–75 mm monopolar needle electrode is used as the cathode. To access the L2/L3/L4 nerve roots, the needle is inserted 2–2.5 cm lateral to the spinous process of the L4 vertebral body. The needle is positioned on the periosteum of the vertebral arch overlying the L4 root. The anode is also a needle electrode and is located on the contralateral side in a similar position (1,2).

To access the L5/S1 nerve roots, a similar setup is used but the needle electrodes are inserted just medial and a bit caudal to the posterior superior iliac spine (1,2).

An alternate technique for the L5/S1 root has been described, which involves placing a superficial electrode on the abdomen as the anode. It is placed opposite the cathode (3).

Stimulation point 2 (S2) (optional): For the L2/L3/L4 nerve roots a second stimulus can be applied to the femoral nerve at the inguinal region (see section on femoral nerve). For the L5/S1 roots a second stimulus can be applied to the sciatic nerve at the gluteal fold (see section on sciatic nerve). The latency from S2 stimulation is subtracted from the S1 latency to calculate a trans-plexus conduction time (1,2).

Machine settings: Sensitivity—2-5 mV/division, Low frequency filter—10 Hz, High frequency filter—10 kHz, Sweep speed—2-5 msec/division (1).

Normal values:

L5/S1 latency to the soleus (msec) (3)

Mean	S.D.
15.4	1.3

Side to side difference: 0.2 msec (0.0–0.8)

Latency to tibialis anterior (msec) (4) (12 subjects)

Mean	S.D.	Range
13.5	1.2	11.4–15.9

Latency to tibialis anterior (msec) (5) (30 subjects)

Mean	S.D.	Range
12.4	1.3	10.8–13.7

Side to side difference: mean 0.3 ± 0.2 msec (range 0.0–0.9)
 Upper limit of normal difference: < 0.7 msec

Latency to flexor hallucis brevis (msec) (4) (14 subjects)

Mean	S.D.	Range
25.1	2.0	21.7–29.7

L2/L3/L4 Trans-plexus conduction time recording from the vastus medialis (msec) (1,2)

Mean	S.D.	Mean + 2 S.D.	Range
3.4	0.6	4.6	2.0–4.4

Side to side difference: 0.0–0.9 msec

L5/S1 Trans-plexus conduction time recording from the abductor hallucis (msec) (1,2)

Mean	S.D.	Mean + 2 S.D.	Range
3.9	0.7	5.3	2.5-4.9

Side to side difference: 0.0–1.0 msec

Amplitude to tibialis anterior (mV) (5) (30 subjects)

Mean	S.D.	Range
5.7	2.4	3.2–10.5

Side to side difference: mean 3.8% ± 2.9 (range 1.4–12.7)
Upper limit of normal difference: 9.6%

Area to tibialis anterior (µVsec) (5) (30 subjects)

Mean	S.D.	Range
25.2	9.6	15.7–41.3

Side to side difference: 6.1% ± 3.1 (range 3.6–17.1)
Upper limit of normal difference: 12.3%

Helpful Hints

- A reduction in amplitude of 50% or more between limb stimulation and root stimulation has been used to define a proximal conduction block. The limb stimulation is applied above the common sites of nerve compression (for example, for the peroneal nerve it is applied above the knee) (6).

Notes

REFERENCES

1. Dumitru: D: *Electrodiagnostic medicine.* Philadelphia: Hanley and Belfus, 1995.
2. MacLean IC: Spinal nerve stimulation. In AAEM Course B: nerve conduction studies—a review. American Association of Electrodiagnostic Medicine, Rochester, Minnesota, 1988.
3. Kraft GH, Johnson EW: Proximal motor nerve conduction and late responses: an AAEM workshop. American Association of Electrodiagnostic Medicine, 1986.
4. Macdonnell RAL, Cros D, Shahani BT: Lumbosacral nerve root stimulation comparing electrical with surface magnetic coil techniques. *Muscle Nerve* 1992; 15:885–890.
5. Chang CW, Lien IN: Spinal nerve stimulation in the diagnosis of lumbosacral radiculopathy. *Am J Phys Med Rehabil* 1990; 69:318–322.
6. Menkes DL, Hood DC, Ballesteros RA, Williams DA: Root stimulation improves the detection of acquired demyelinating polyneuropathies. *Muscle Nerve* 1988; 21:298–308.

ADDITIONAL READING/ALTERNATE TECHNIQUE

1. Troni W, Bianco C, Moja MC, Dotta M: Improved methodology for lumbosacral nerve root stimulation. *Muscle Nerve* 1996; 19:595–604.

CHAPTER 7

Other Studies of Interest

The following is a list of some other nerve conduction tests that have been published. This list obviously is not all-inclusive, but it does give the reader references for other studies that may be necessary from time to time. It is anticipated that most of these studies will not be performed routinely by most electrodiagnosticians. As a rule, these studies also require more in-depth reading, study, and practice than is possible to summarize in the format of the preceding sections.

Autonomic Nervous System

1. Ravits JM: AAEM minimonograph #48: autonomic nervous system testing. *Muscle Nerve* 1997; 20:919–937.

Axillary F-Loop/Central Latency

1. Hong CZ, Joynt RL, Lin JC, et al: Axillary F-loop latency of ulnar nerve in normal young adults. *Arch Phys Med Rehabil* 1981; 62:565–569.
2. Wu Y, Kunz JRM, Putnam TD, Stratigos JS: Axillary F central latency: simple electrodiagnostic technique for proximal neuropathy. *Arch Phys Med Rehabil* 1983; 64:117–120.

Axillary Sensory Nerve Conduction

Karandreas N, Papatheodorou A, Triantaphilos I, et al: Sensory nerve conduction studies of the less frequently examined nerves. *Electromyogr clin Neurophysiol* 1995; 35:169–173.

Bulbocavernosus Reflex/Perineal/Pudendal Nerve Conduction

1. Dick HC, Bradley WE, Scott FB, Timm GW: Pudendal sexual reflexes: electrophysiologic investigations. *Urology* 1974; 3:376–379.
2. Siroky MB, Sax DS, Krane RJ: Sacral signal tracing: the electrophysiology of the bulbocavernosus reflex. *J Urol* 1979; 122:661–664.
3. Kiff ES, Swash M: Slowed conduction in the pudendal nerves in idiopathic (neurogenic) faecal incontinence. *Br J Surg* 1984; 71:614–616.
4. Snooks SJ, Swash M: Perineal nerve and transcutaneous spinal stimulation: new methods for investigation of the urethral striated sphincter musculature. *Br J Urol* 1984; 56:406–409.

Other Studies of Interest **275**

5. Tetzschner T, Sorensen M, Lose G, Christiansen J: Vaginal pudendal nerve stimulation: a new technique for assessment of pudendal nerve terminal motor latency. *Acta Obstet Gynecol Scand* 1997; 76:294–299.

Dorsal Nerve of the Penis

1. Bradley WE, Lin JTY, Johnson B: Measurement of the conduction velocity of the dorsal nerve of the penis. *J Urol* 1984; 131:1127–1129.
2. Clawson DR, Cardenas DD: Dorsal nerve of the penis nerve conduction velocity: a new technique. *Muscle Nerve* 1991; 14:845–849.
3. Herbaut AG, Sattar AA, Salpigides G, et al: Sensory conduction velocity of dorsal nerve of the penis during pharmacoerection: a more physiological technique? *Eur Urol* 1996; 30:60–64.

Dorsal Rami, Sensory Conduction

1. Singh AP, Sommer HM: Sensory nerve conduction studies of the L-1/L-2 dorsal rami. *Arch Phys Med Rehabil* 1996; 77:913–915.

Dorsal Scapular Nerve Motor Conduction

1. LoMonaco M, DiPasqua RG, Tonali P: Conduction studies along the accessory, long thoracic, dorsal scapular, and thoracodorsal nerves. *Acta Neurol Scand* 1983; 68:171–176.

Greater Auricular Sensory Nerve Conduction

1. Palliyath SK: A technique for studying the greater auricular nerve conduction velocity. *Muscle Nerve* 1984; 7:232–234.
2. Kimura I, Seki H, Sasao SI, Ayyar DR: The greater auricular nerve conduction study: a technique, normative data and clinical usefulness. *Electromyogr clin Neurophysiol* 1987; 27:39–43.

Hypoglossal Motor Nerve Conduction

1. Redmond MD, DiBenedetto M: Electrodiagnostic evaluation of the hypoglossal nerve. *Arch Phys Med Rehabil* 1984; 65:633.
2. Redmond MD, DiBenedetto M: Hypoglossal nerve conduction in normal subjects. *Muscle Nerve* 1988; 11:447–452.

Ilioinguinal Nerve Conduction

1. Ellis RJ, Geisse H, Holub BA, et al.: Ilioinguinal nerve conduction. *Muscle Nerve* 1992; 15:1194.

Intercostal Nerve Conduction

1. Caldwell JW, Crane CR, Boland GL: Determinations of intercostal motor conduction time in diagnosis of nerve root compression. *Arch Phys Med Rehabil* 1968; 49:515–518.
2. Pradhan S, Taly A: Intercostal nerve conduction study in man. *J Neurol Neurosurg Psychiatry* 1989; 52:763–766.

Posterior Interosseous Sensory Response

1. Pelier-Cady MC, Raimbeau G, Saint Cast Y: Posterior interosseous nerve: sensory nerve conduction technique. Presented at the AAEM annual meeting, September 20, 1997.

Quadriceps Late Responses

1. Mongia SK: H reflex from quadriceps and gastrocnemius muscles. *Electromyography* 1972; 12:179–190.
2. Aiello I, Serra G, Rosati G, Tugnoli V: A quantitative method to analyze the H reflex latencies from vastus medialis muscle: normal values. *Electromyogr clin Neurophysiol* 1982; 22:251–254.
3. Kameyama O, Hayes KC, Wolfe D: Methodological considerations contributing to variability of the quadriceps H-reflex. *Am J Phys Med Rehabil* 1989; 68:277–282.
4. Garland SJ, Gerilovsky L, Enoka RM: Association between muscle architecture and quadriceps femoris H-reflex. *Muscle Nerve* 1994; 17:581–592.
5. Wochnik-Dyjas D, Glazowski C, Niewiadoska M: The F-wave in the vastus lateralis M. and the segmental motor conduction times for L2/L4 motoneurons. *Electroencephalogr clin Neurophysiol* 1996; 101:379–386.

Radial Nerve Motor Conduction Study to the Triceps

1. Gassel MM: A test of nerve conduction to muscles of the shoulder girdle as an aid in the diagnosis of proximal neurogenic and muscular disease. *J Neurol Neurosurg Psychiatry* 1964; 27:200–205.

Repetitive Stimulation

1. Litchy WJ, Albers JW: Repetitive stimulation, an AAEM workshop. American Association of Electrodiagnostic Medicine, Rochester, Minnesota, 1984.
2. Keesey JC: AAEM minimonograph #33: electrodiagnostic approach to defects of neuromuscular transmission. *Muscle Nerve* 1989; 12:613–626.

Trigeminal Motor Nerve Conduction

1. Dillingham TR, Spellman NT, Chang AS: Trigeminal motor nerve conduction: deep temporal and mylohyoid nerves. *Muscle Nerve* 1996; 19:277–284.

Trigeminal Sensory Nerve Conduction

1. Raffaele R, Emery P, Palmeri A, et al: Sensory conduction velocity of the trigeminal nerve. *Electromyogr clin Neurophysiol* 1987; 27:115–117.

APPENDIX

BMI Tables

BMI (body mass index—kg/m^2)

Metric Calculation

Weight (kg)

Height (cm)	45	50	55	60	65	70	75	80	85	90	95	100
150	20	22	24	27	29	31	33	36	38	40	42	44
160	18	20	21	23	25	27	29	31	33	35	37	39
170	16	17	19	21	22	24	26	28	29	31	33	35
180	14	15	17	19	20	22	23	25	26	28	29	31
190	12	14	15	17	18	19	21	22	24	25	26	28
200	11	13	14	15	16	18	19	20	21	23	24	25

English Calculation

Weight (lbs)

Height (in)	110	115	120	125	130	135	140	145	150	155	160	165	170	175	180	185	190	195	200	205	210	215	220	225	230	235
5'0"	21	22	23	24	25	26	27	28	29	30	31	32	33	34	35	36	37	38	39	40	41	42	43	44	45	46
5'1"	21	22	23	24	25	26	27	27	28	29	30	31	32	33	34	35	36	37	38	39	40	41	42	43	43	44
5'2"	20	21	22	23	24	25	26	27	27	28	29	30	31	32	33	34	35	36	37	37	38	39	40	41	42	43
5'3"	19	20	21	22	23	24	25	26	27	27	28	29	30	31	32	33	34	35	35	36	37	38	39	40	41	42
5'4"	19	20	21	21	22	23	24	25	26	27	27	28	29	30	31	32	33	33	34	35	36	37	38	39	39	40
5'5"	18	19	20	21	22	22	23	24	25	26	27	27	28	29	30	31	32	32	33	34	35	36	37	37	38	39
5'6"	18	19	19	20	21	22	23	23	24	25	26	27	27	28	29	30	31	31	32	33	34	35	36	36	37	38
5'7"	17	18	19	20	20	21	22	23	23	24	25	26	27	27	28	29	30	31	31	32	33	34	34	35	36	37
5'8"	17	17	18	19	20	21	21	22	23	24	24	25	26	27	27	28	29	30	30	31	32	33	33	34	35	36
5'9"	16	17	18	18	19	20	21	21	22	23	24	24	25	26	27	27	28	29	30	30	31	32	33	33	34	35
5'10"	16	17	17	18	19	19	20	21	22	22	23	24	24	25	26	27	27	28	29	29	30	31	32	32	33	34
5'11"	15	16	17	17	18	19	20	20	21	22	22	23	24	24	25	26	26	27	28	29	29	30	31	31	32	33
6'0"	15	16	16	17	18	18	19	20	20	21	22	22	23	24	24	25	26	26	27	28	28	29	30	31	31	32
6'1"	15	15	16	16	17	18	18	19	20	20	21	22	22	23	24	24	25	26	26	27	28	28	29	30	30	31
6'2"	14	15	15	16	17	17	18	19	19	20	21	21	22	22	23	24	24	25	26	26	27	28	28	29	30	30
6'3"	14	14	15	16	16	17	17	18	19	19	20	21	21	22	22	23	24	24	25	26	26	27	27	28	29	29
6'4"	13	14	15	15	16	16	17	18	18	19	19	20	21	21	22	23	23	24	24	25	26	26	27	27	28	29

Index

284 Index

Axillary motor nerve to the deltoid, 2–5

Blink reflex, 250–255

Cervical nerve root stimulation, 264–267
Comparative upper extremity sensory and mixed studies, 138–152
Cranial nerve VII, 256–259
Cranial nerve XI, 260–262
Cranial nerves, 249–262   blink reflex, 250–255
  cranial nerve VII, 256–259
  cranial nerve XI, 260–262

Deep peroneal sensory study, 214–217

Femoral motor nerve to the quadriceps, 154–157

H–reflex to the calf, 194–198

Lateral antebrachial cutaneous sensory nerve study, 92–97
Lateral femoral cutaneous sensory study, 200–205
  Ma and Liveson technique, 200–201
  Spevak and Prevec technique, 202–204
Long thoracic nerve motor study to the serratus anterior, 6–9
Lower extremity motor nerves, 153–198
  femoral motor nerve to the quadriceps, 154–157
  peroneal nerve, 158–175
    peroneal motor nerve to the extensor digitorum brevis, 158–163
    peroneal motor nerve to the peroneus brevis, 164–167

peroneal motor nerve to the peroneus longus, 168–171
peroneal motor study to the tibialis anterior, 172–175
sciatic nerve, 176–179
tibial nerve, 180–193
  tibial motor nerve (inferior calcaneal branch) to the abductor digiti minimi, 180–183
  tibial motor nerve (medial plantar branch) to the abductor hallucis, 184–189
  tibial motor nerve (lateral plantar branch) to the flexor digiti minimi brevis, 190–193
H–reflex to the calf, 194–198

Lower extremity sensory and mixed studies, 199–248
  lateral femoral cutaneous sensory study, 200–205
    Ma and Liveson technique, 200–201
    Spevak and Prevec technique, 202–204
  medial calcaneal sensory study, 206–209
  medial femoral cutaneous sensory study, 210–213
  peroneal nerve, 214–227
    deep peroneal sensory study, 214–217
    superficial peroneal sensory study (medial and intermediate dorsal cutaneous branches), 218–223
    posterior femoral cutaneous sensory study, 224–227
  saphenous nerve, 228–235
    sensory study (distal technique), 228–231

Index   285

saphenous sensory study (proximal technique), 232–235
sural nerve, 236–243
  sural lateral dorsal cutaneous branch sensory study, 236–239
  sural sensory study, 240–243
  tibial nerve compound action potentials (medial and lateral plantar nerves), 244–248
Lumbar root stimulation, 268–272

Medial antebrachial cutaneous sensory nerve study, 98–103
Medial calcaneal sensory study, 206–209
Medial femoral cutaneous sensory study, 210–213
Median and radial sensory nerves to 1st digit, 138–141
Median and ulnar mixed nerve studies, 142–147
Median and ulnar sensory studies to the 4th digit, 148–152
Median motor nerve (anterior interosseous branch) conduction to the flexor pollicis longus, 24–27
Median motor nerve (anterior interosseous branch) to the pronator quadratus, 28–31
Median motor nerve to the 1st lumbrical, 32–35
Median motor nerve to the 2nd lumbrical, 36–39
Median nerve to the abductor pollicis brevis, 10–17
Median nerve to the flexor carpi radialis, including H–reflex, 18–23
Median palmar cutaneous sensory nerve study, 112–115

Median sensory nerve to the 2nd and 3rd digits, 104–111
Musculocutaneous motor nerve to the biceps brachii, 40–43

Other studies of interest, 273–277
  autonomic nervous system, 274
  axillary F–loop/central latency, 274
  axillary sensory nerve conduction, 174
  bulbocavernosus reflex/perineal/pudendal nerve conduction, 274–275
  dorsal nerve of the penis, 275
  dorsal rami, sensory conduction, 275
  dorsal scapular nerve motor conduction, 275
  greater auricular sensory nerve conduction, 275
  hypoglossal motor nerve conduction, 275
  ilioinguinal nerve conduction, 276
  intercostal nerve conduction, 276
  posterior interosseous sensory response, 276
  quadriceps late responses, 276
  radial nerve motor conduction study to the triceps, 276
  repetitive stimulation, 277
  trigeminal motor nerve conduction, 277
  trigeminal sensory nerve conduction, 277

Peroneal motor nerve to the extensor digitorum brevis, 158–163
Peroneal motor nerve to the peroneus brevis, 164–167
Peroneal motor nerve to the peroneus longus, 168–171
Peroneal motor study to the tibialis anterior, 172–175

Index

Phrenic motor nerve to the diaphragm, 44–49
Posterior antebrachial cutaneous sensory nerve study, 116–119
Posterior femoral cutaneous sensory study, 224–227

Radial motor nerve to the extensor digitorum communis, 54–57
Radial motor nerve to the extensor indicis proprius, 58–63
Radial (posterior interosseous) motor nerve to the extensor carpi ulnaris and brachioradialis muscles, 50–53
Radial sensory nerve study to the base of the thumb, 120–123
Root stimulation, 263–272
  cervical nerve root stimulation, 264–267
  lumbar root stimulation, 268–272

Saphenous nerve, 228–235
  distal technique, 228–231
  proximal technique, 232–235
Sciatic nerve, 176–179
Superficial peroneal sensory study (medial and intermediate dorsal cutaneous branches), 218–223
Suprascapular motor nerve to the supraspinatus and infraspinatus, 64–69
Sural lateral dorsal cutaneous branch sensory study, 236–239
Sural sensory study, 240–243

Thoracodorsal motor nerve to the latissimus dorsi, 70–73
Tibial motor nerve (inferior calcaneal branch) to the abductor digiti minimi, 180–183

Tibial motor nerve (lateral plantar branch) to the flexor digiti minimi brevis, 190–193
Tibial motor nerve (medial plantar branch) to the abductor hallucis, 184–189
Tibial nerve compound action potentials (medial and lateral plantar nerves), 244–248

Ulnar dorsal cutaneous sensory nerve study, 124–129
Ulnar motor nerve to the 1st dorsal interosseous, 86–90
Ulnar motor nerve to the anterior digiti minimi, 74–81
Ulnar motor nerve to the palmar interosseous muscle, 82–85
Ulnar sensory nerve to the 5th digit, 130–137
Upper extremity sensory and mixed nerves, 91–152
  lateral antebrachial cutaneous sensory nerve study, 92–97
  medial antebrachial cutaneous sensory nerve study, 98–103
  median nerve, 104–123
    median sensory nerve to the 2nd and 3rd digits, 104–111
    median palmar cutaneous sensory nerve study, 112–115
    posterior antebrachial cutaneous sensory nerve study, 116–119
    radial sensory nerve study to the base of the thumb, 120–123
  ulnar nerve, 124–137
    ulnar dorsal cutaneous sensory nerve study, 124–129
    ulnar sensory nerve to the 5th digit, 130–137
  comparative studies, 138–152

median and radial sensory
  nerves to the 1st digit,
  138–141
median and ulnar mixed nerve
  studies, 142–147
median and ulnar sensory
  studies to the 4th digit,
  148–152
Upper extremity/cervical
  plexus/brachial plexus motor
  studies, 1–90
axillary motor nerve to the deltoid,
  2–5
long thoracic nerve motor study to
  the serratus anterior, 6–9
median nerve, 10–43
  median motor nerve to the
    abductor pollicis brevis,
    10–17
  median nerve to the flexor carpi
    radialis, including H–reflex,
    18–23
  median motor nerve (anterior
    interosseous branch)
    conduction to the flexor
    pollicis longus, 24–27
  median motor nerve (anterior
    interosseous branch) to the
    pronator quadratus, 28–31
  median motor nerve to the 1st
    lumbrical, 32–35
  median motor nerve to the 2nd
    lumbrical, 36–39
musculocutaneous motor nerve to
  the biceps brachii, 40–43
phrenic motor nerve to the
  diaphragm, 44–49
radial nerve, 50–63
  radial (posterior interosseous)
    motor nerve to the extensor
    carpi ulnaris and
    brachioradialis muscles,
    50–53
  radial motor nerve to the
    extensor digitorum
    communis, 54–57
  radial motor nerve to the
    extensor indicis proprius,
    58–63
suprascapular motor nerve to the
  supraspinatus and
  infraspinatus, 64–69
thoracodorsal motor nerve to the
  latissimus dorsi, 70–73
ulnar nerve, 74–90
  ulnar motor nerve to the
    abductor digiti minimi,
    74–81
  ulnar motor nerve to the palmar
    interosseous muscle, 82–85
  ulnar motor nerve to the 1st
    dorsal interosseous, 86–90